Combinatorial Complexes

Mathematics and Its Applications

Managing Editor:

M. HAZEWINKEL
Department for Mathematics, Erasmus University, Rotterdam, The Netherlands

Editorial Board:

R. W. BROCKETT. *Harvard University, Cambridge, Mass., U.S.A.*
Yu. I. MANIN, *Steklov Institute of Mathematics, Moscow, U.S.S.R.*
G.-C. ROTA, *M.I.T., Cambridge, Mass. U.S.A.*

Volume 2

Peter H. Sellers

Senior Research Associate
The Rockefeller University, New York

Combinatorial Complexes

A Mathematical Theory of Algorithms

D. REIDEL PUBLISHING COMPANY

Dordrecht : Holland / Boston : U.S.A. / London : England

Library of Congress Cataloging in Publication Data

Sellers, Peter H.
     Combinatorial complexes.

     (Mathematics and its applications; v. 2)
     Bibliography: p.
     Includes index.
     1.   Electronic digital computers–Programming.   2.   Programming languages
(Electronic computers)   3.   Algorithms.   I.   Title.   II.   Series: Mathematics
and its applications (Dordrecht); v. 2.
QA76.6.S448   001.6'42   79-16329

Published by D. Reidel Publishing Company
P.O. Box 17, Dordrecht, Holland

Sold and distributed in the U.S.A., Canada, and Mexico
by D. Reidel Publishing Company, Inc.
Lincoln Building, 160 Old Derby Street, Hingham, Mass. 02043, U.S.A.

Printed in The Netherlands

To *Lucy Bell*

Editor's Preface

Approach your problems from the right end and begin with the answers. Then, one day, perhaps you will find the final question.

'The Hermit Clad in Crane Feathers' in R. Van Gulik's *The Chinese Maze Murders*.

It isn't that they can't see the solution. It is that they can't see the problem.

G. K. Chesterton, The scandal of Father Brown "The point of a pin"

Growing specialization and diversification have brought a host of monographs and textbooks on increasingly specialized topics. However, the 'tree' of knowledge of mathematics and related fields does not grow only by putting forth new branches. It also happens, quite often in fact, that branches which were thought to be completely disparate are suddenly seen to be related.

Further, the kind and level of sophistication of mathematics applied in various sciences has changed drastically in recent years: measure theory is used (non-trivially) in regional and theoretical economics; algebraic geometry interacts with physics; the Minkowsky lemma, coding theory and the structure of water meet one another in packing and covering theory; quantum fields, crystal defects and mathematical programming profit from homotopy theory; Lie algebras are relevant to filtering; and prediction and electrical engineering can use Stein spaces.

This series of books, *Mathematics and Its Applications*, is devoted to such (new) interrelations as *exempla gratia*:

- a central concept which plays an important role in several different mathematical and/or scientific specialized areas;
- new applications of the results and ideas from one area of scientific endeavor into another;
- influences which the results, problems and concepts of one field of enquiry have and have had on the development of another.

With books on topics such as these, of moderate length and price, which are stimulating rather than definitive, intriguing rather than encyclopaedic, we hope to contribute something towards better communication among the practitioners in diversified fields.

The unreasonable effectiveness of mathematics in science . . .

> Eugene Wigner

Well, if you knows of a better 'ole, go to it.

> Bruce Bairnsfather

What is now proved was once only imagined.

> William Blake

As long as algebra and geometry proceeded along separate paths, their advance was slow and their applications limited.

But when these sciences joined company, they drew from each other fresh vitality and thenceforward marched on at a rapid pace towards perfection.

> Joseph Louis Lagrange

Krimpen a/d IJssel
March, 1979.

Michiel Hazewinkel

Table of Contents

Preface

The principal object of this book is to introduce a workable mathematical theory, aimed at developing new algorithms. This is a departure from the usual mathematical approach to algorithms, which is to analyze known ones, as is done, for instance, in Donald E. Knuth's definitive book, *The Art of Computer Programming* [4], which describes itself as being on "the theory of the properties of particular computer algorithms". As long as a theory is confined to particular algorithms, there is no need to establish the limits of what would be considered as an algorithm. However, if our object is to search for new algorithms, we need to define the area in which the search will take place.

An algorithm is a special kind of theorem, rather than a mathematically defined structure. Therefore, to reduce the search for an algorithm to mathematical terms, it can be regarded as a search for a structure which represents an algorithm, as faithfully as possible. The usual way to represent an algorithm is by a program, written in a given language. However, there can be many programs for a single algorithm, because an algorithm is largely independent of the language in which it is written, and, even in one language, it can be represented by more than one program. Therefore, a more faithful way to represent it is by an equivalence class of programs. It is also a simpler way, because the algorithm is represented in effect by the invariants of that class of programs.

My point of departure for the development of these ideas into the mathematical theory is to define a *programming language*, as a basis for a chain complex, in the same way that a topological space is described in algebraic topology. Then a *language isomorphism* is chain isomorphism, and two programs are defined as being *equivalent*, when they are joined by such an isomorphism. This is how the equivalence class of programs is determined, which is used to represent an algorithm. Chapter I of this book uses this method of representing algorithms in the development of a theory which is aimed at finding new algorithms. The subsequent chapters are applications of the theory, largely independent of each other, which demonstrate its use as a mathematical tool.

A minimum selection of chapters to read would be the Introduction, Chapter I, and one other chapter, say Chapter V, which is a short one. The intervening chapters are more complicated, because originally each one was a pair of related chapters, which have been merged into one, having the original ones as special cases. Chapters V and VI can probably be merged in the same way, but it has not been done, which means that they are conceptually simpler than the others, and easier to read.

The Rockefeller University   PETER H. SELLERS
New York
December, 1978

Acknowledgements

I am pleased to have this opportunity of thanking those who have helped me toward the publication of this book: Gian-Carlo Rota for inviting me to contribute it to the MIA series; Mark Kac for encouraging my long research into the remote areas in which the discoveries of this book were made; Britton Chance, Herbert Wilf, John Riordan, Morris Schreiber, Mary Ellen O'Brien, Marie Grossi, Julia Gonzalez and Mary Sharkey for their ready help along the way.

Chapter 0

Introduction

0.1. FINDING ALGORITHMS

A mathematical theory of algorithms is introduced here, which has been developed with an aim toward finding new algorithms, rather than analyzing known ones. One important approach to the study of algorithms is based on the idea that each one can be expressed by a program on a Turing Machine or some equivalent model of computation. However, it is not feasible to try to find a desired algorithm by exploring the set of all Turing Machine programs. What is proposed here, instead, is a theory in which an algorithm is expressed as a combination of major computational steps, each of which can be carried out in a manner which is assumed to be known. In other words, an algorithm is subdivided into a relatively small number of parts, so that its global structure may be readily expressible in terms of the incidence relations among them. Such a theory will be developed in chapter 1, and applications of it will be discussed in subsequent chapters.

A comparison of the usual way of talking about algorithms in mathematics with the contents of this book raises some immediate questions:

(i) Why are programs talked about rather than algorithms?

(ii) Why are different languages used to express different algorithms rather than one universal language for all algorithms?

(iii) Why are list-making algorithms preferred?

1

Algorithms are treated in this book in the same way that geometrical figures are treated in algebraic topology. An algorithm is subdivided into a family of subalgorithms, and its properties are deduced from the incidence relations among them. The subalgorithms are regarded as irreducible, and each one is represented by a single element, called an *instruction*. The whole algorithm is represented by a set, including such instructions and also input data. There are incidence relations among the elements of this set, because each instruction depends on and produces other elements of the set, which are said to be *incident* to it and together are described as its *boundary*.

Before undertaking to represent a particular algorithm in the above way, we must define a suitable set of data and instructions with incidence relations among them. The *data* are those elements of the set which have no elements incident to them, that is, no boundary elements, whereas *instructions* always produce and usually depend on other elements, known as their boundary elements. The incidence or boundary relations among elements of the set have the same formal properties as they would among simplices or other types of cells in algebraic topology, which means in algebraic terms that the set generates a *chain complex*. Consequently, it will be possible to represent an algorithm by an algebraic structure, which will be defined as a *program* and is a substructure of a chain complex. It represents, but does not define, an algorithm, because in general there are many programs representing a given algorithm within the context of a given chain complex.

Therefore, in answer to question (i), chain complexes and programs are the main objects of discussion in our theory, rather than algorithms, which are not viewed as formal structures, but as theorems which determine programs non-uniquely.

As our theory of algorithms resembles algebraic topology, so to a less degree does the Turing Machine theory of algorithms resemble point-set topology. To program an algorithm on a Turing Machine means to represent it by data and instructions of the most elementary conceivable kind, like the representation of a topological space by points and neighborhoods. Whereas, to program an algorithm, as we shall do, in a language especially suited to the computation which the algorithm is aimed at carrying out, means to subdivide it into a relatively small set of data and instructions like the subdivision of a polyhedron in combinatorial or algebraic topology.

The set of all data and instructions, allowed to be used in the representation of an algorithm by a program, is called a *programming language*. There is freedom in the way it may be chosen, just as there is freedom in the way a geometrical figure may be subdivided into simplices or other kinds of cells, and the ease of finding programs in the language depends on this choice. Therefore, the answer to question (ii) is that we choose a language which is particularly suited to the kind of computation which the desired program is expected to perform. This means that the objective, loosely described as 'finding an algorithm', is separated into two parts, first, choosing a suitable language, and then finding a program in it.

A programming language is a limited set of data and instructions, such that each instruction depends on and produces other elements of the language. It contains no such thing as a transfer instruction, governing the order in which instructions are to be performed. In fact, the instructions in a program are not necessarily in a linear order, but they are only partially ordered, in such a way that, if an instruction depends on certain elements, not given as input, then it must be preceded by an instruction which produces them. Therefore, the ordering of instructions in a program is determined by the boundary elements (input and output) of each instruction. The fundamental fact remains that in our theory of algorithms all we need to know about an instruction, beyond the fact that it exists, is its boundary.

If a language has been given, then to say that a certain algorithm is desired means that it must produce as output some elements of the language, defined in advance. In general, a minimal number of elements of the language will have to be given as input to the algorithm. Now, suppose we have developed a theory which helps us to find a representative program for any such algorithm, then the best test of the theory will be to find a program which is capable of producing any element of the entire language. This is what a *list-making* program does, and it is evident that other programs, only requiring partial lists of the language, are likely to be derivable from it.

As an example of the subsidiary role of other types of programs, consider the input requirements. The input necessary for any program in a language is a subset of the input necessary for one of the programs which lists without repetition every element of the language. The minimum number of input elements for such a list-making program will be shown to be equal to the rank of the 'homology group' of the complex, generated by the language. This is one of the fundamental results of the theory. Also, it will be shown that every program, which lists without repetitions all the elements of the language, corresponds to what is described in homology theory as a 'standard basis' for the chain complex of the language.

Therefore, in answer to question (iii) list-making programs are preferred, not only because they fit naturally into a theoretical framework for algorithms, based on chain complexes, but also because they make a maximum use of the programming language. This means that other kinds of algorithms are in some sense subsidiary to them.

The main fact about the kind of list-making program under consideration is that any element of its language is listed after a finite number of steps, and it is never listed again in any subsequent step. This implies that the language is a finite or countable set, but here an upper limit will always be placed on a language, so that it may be kept finite. Consequently, we may say that a list-making program generates every element of the language except those which were given in the first place as input. A program which is required to generate only a partial list of the elements of the language can be obtained by leaving some elements off the list, or keeping them on a separate temporary list, until all the instructions which depend on them have been performed. As an example of this consider the Euclidean Algorithm.

A pair of positive integers is given, whose greatest common divisor is to be found. This can be seen as a list-making program which produces a set of pairs of positive integers, no larger than the ones given, all pairs having the same greatest common divisor. The program stops when a pair of equal numbers is produced. This pair alone is listed. The number, appearing twice in it, is the desired greatest common divisor.

Another example is as follows: Suppose we want to find a sorting algorithm, which will take a finite list and transpose adjacent pairs of elements repeatedly, until the whole list is in some pre-assigned linear ordering. This can be done by a special case of a list-making program, constructed in Chapter IV, which lists without repetitions not only all the permutations of a given finite sequence, but also all the elements of the programming language. It lists all the permutations first, using transposition instructions, and it may be stopped at this point. The only input which it requires up to then is one sequence, whose terms are in non-decreasing order. (The algorithm works regardless of what linear ordering we assign to the set from which the terms of the sequence are taken.)

Suppose now we start with any sequence S, obtainable by permuting the input sequence I. Then our list-making program contains a unique succession of transposition instructions from I to S. Furthermore, the transposition instructions are in one-to-one correspondence with the sequences they produce, and given any intermediate sequence between I and S, we know the transposition instruction which produced it. Therefore, starting with S and applying the inverse of the transposition, which produced it, to S, we get a new sequence. Again we apply the inverse of the transposition, which produced it, and we continue thus until I is reached. This is a sorting algorithm, and any sorting algorithm may be so obtained from some list-making algorithm for permutations.

Having suggested the need for a global theory of algorithms, which would ignore their fine structure, as expressed in the theory of Turing Machines, and, having answered some basic questions about the form of the theory, let us continue this introduction with a preview of the way in which the theory is developed.

0.2. PROGRAMS REPRESENT ALGORITHMS

There seems to be no generally accepted rule as to how much must be known about a program before it can be regarded as an adequate characterization of an algorithm. Presumably we would have to know which features of the program belong to the algorithm, and which are merely manifestations of the way in which the program was written. It would be ideal if we could define the set of all programs which represent a single algorithm, and then find a set of invariants of these programs, by which to characterize the underlying algorithm. This is the direction taken in this book. However, an algorithm will be represented, not by all possible programs, but by a limited set of programs, which is readily definable as an isomorphism class relative to a few simple axioms. Two programs are isomorphic, if they are joined by a one-to-one correspondence which preserves the axioms, though at the level of computation isomorphic programs may be very different. Thus, it becomes possible to find a program in two steps: First, find one which is conceptually simple, regardless of whether it has anything to recommend it from a computational viewpoint, and, secondly, apply a succession of isomorphisms to it, until it takes a form which is suited to whatever computational tools we plan to use.

Programs and programming languages will be defined and a theory of programs developed in Chapter I with a view toward finding explicit programs in the subsequent chapters. Each chapter after the first is concerned with a finite set of mathematical interest, and the initial object of the chapter will be to construct a program which lists every element of the set without repetitions. Whatever programming language is used, will include this set. The final object of the chapter will be to find a program which lists every element of the entire language without repetitions. There will be three principal steps in finding such a program:

(i) Choose a language. That is, decide what instructions are allowed to be used, and symbolize them in a way which makes it clear what elements of the language each instruction (also in the language) depends on and produces.

(ii) Find a program. This reduces to a mathematical problem and is the principal concern of this book. Finding a program will take the form of finding a basis for a finitely generated Abelian group.

(iii) Specify the manner in which the program is to be carried out. This must take into account the computational tools to be used, but it does not alter the underlying program. This step goes beyond the scope of this book, since we are not conducting a study of the tools by which programs are carried out — computers, for instance.

A standard way of 'specifying a program' will be adopted for the purposes of this book. It will perform this step by placing the instructions of the program in a particular linear ordering, but it must be remembered that this is only a sample method. Obviously, a program does not have to be carried out sequentially. Portions of it could be executed in 'parallel'.

The first two steps, choosing a language and finding a program in it, are our principal topics. We do not follow the method, ordinarily used in studies of algorithms, of adopting a universal language, in which it is presumed that any algorithm can be expressed by a suitable program, but rather, we tailor the language to the problem so that it will be as easy as possible to find a program in it. It does not matter whether the instructions in the language are convenient for computation, or not, so long as we can find a program, using them, which carries out the desired overall computation. After that, by a succession of language isomorphisms we can transform the first-found program into one which is more convenient to apply.

The final development of a program, specifying the manner in which it is to be carried out, is deemed to be independent of the program itself. According to our definition, a program is completely determined without regard to the question of how it could be reduced to a flowchart or to some non-sequential procedure for carrying it out. In general, there are many such procedures, sequential or not, which are regarded merely as options, available to a single program.

0.3. PROGRAMMING LANGUAGES

A programming language is a set of data and instructions, bound by a simple axiomatic structure. Each instruction stands for an explicit computation, which depends on some elements of the language and produces others. For example we could define an integer multiplication language to include, as data, every integer and, as instructions, every ordered pair $i \times j$ of integers, where $i \times j$ is the instruction which tells us to multiply i times j. This instruction is explicit in the sense that it tells us which two integers to multiply, what their product is, and in what order to multiply them. It is said to *depend on* the two integers i and j, and to *produce* the integer ij. In this language, if $i \neq j$, then $i \times j$ and $j \times i$ are distinct instructions, even though they depend on and produce the same elements. The language could include, if we wanted it to, a transposition instruction which would depend on $i \times j$ and produce $j \times i$, which would be an example of how an instruction can act upon other instructions.

Let us require programming languages in general to satisfy two axioms. The first assigns an integral dimension to each element of the language, such that, if π is the dimension of an instruction, then $\pi - 1$ is the dimension of the elements it depends on and produces. In the sample language each integer may be given dimension 1, each $i \times j$ dimension 2, and the instruction which converts $i \times j$ into $j \times i$ dimension 3.

The second axiom is easiest to state in the case of a language where every instruction p depends on exactly one element q and produces exactly one element r. In other words, p converts q into r. Then, it is required for any two instructions q and r, for which there exists p which converts one into the other, that q and r be equivalent, in the sense that they both depend on the same element and both produce the same element. Thus, the result of performing p is to replace q by another instruction r, which has exactly the same computational effect as q. In the sample language the instruction which converts $i \times j$ into $j \times i$, satisfies the axiom, since $i \times j$ is equivalent to $j \times i$. Such a transposition instruction would obey the second axiom in any commutative multiplication language.

The second axiom is extended to languages whose instructions depend on or produce more than one element by introducing formal sums, by means of which several elements can be treated as one. For instance, the instruction $i \times j$ depends on the formal sum $i + j$ and produces ij. The plus sign is that of a free Abelian group, not that of the integers. Our procedure, in general, is to embed each language in the free Abelian group which it generates and extend language isomorphisms to the unique linear transformation which they determine. Then the second axiom can be stated as follows:

If p is an instruction, let

$$\Delta(p) = -x + y,$$

where p depends on x and produces y, each being a formal sum, then

$$\Delta(x) = \Delta(y).$$

If p is not an instruction, let

$$\Delta(p) = 0.$$

Regarding Δ as a linear transformation, allows us to express the axiom by

$$\Delta\Delta = 0,$$

since for any p

$$\Delta(\Delta(p)) = \Delta(-x + y) = -\Delta(x) + \Delta(y) = 0.$$

Let us return to our sample language and define a new instruction $i \times j \times k$, which depends on two instructions,

$$j \times k + i \times jk$$

and produces two instructions

$$i \times j + ij \times k.$$

Hence,

$$\Delta(i \times j \times k) = -(j \times k + i \times jk) + (i \times j + ij \times k).$$

Applying Δ to both sides, we get

$$\Delta(\Delta(i \times j \times k)) = 0,$$

which shows that $i \times j \times k$ conforms with the second axiom. This formula is merely a restatement of the associative law for integer multiplication. Hence, $i \times j \times k$ would satisfy the second axiom in any associative multiplication language.

What we have done is to define a language in such a way that it would be expressible by a *chain complex*, i.e. a sequence

$$\ldots \overset{\Delta}{\to} C_\pi \overset{\Delta}{\to} C_{\pi-1} \overset{\Delta}{\to} \ldots$$

of free Abelian groups, where the free generators of C_π are the instructions of dimension π, and Δ is such as to
(i) lower dimensions by 1, and
(ii) satisfy $\Delta\Delta = 0$.
These are the two axioms.

Therefore, just as is done with geometrical figures in algebraic topology, a programming language is embedded in a chain complex

$$\ldots \overset{\Delta}{\to} C_3 \overset{\Delta}{\to} C_2 \overset{\Delta}{\to} C_1 \overset{\Delta}{\to} 0$$

which exhibits its global structure. This chain complex may be expressed more briefly by (C, Δ), where C the direct sum

$$C = C_1 + C_2 + C_3 + \ldots,$$

and Δ becomes extended to a linear transformation of C into itself.

0.4. LIST-MAKING ALGORITHMS

Once a programming language has been chosen, the kind of list-making algorithm we look for to list the entire language without repetitions and with a minimum size of input, is of special interest, not only because of its all-inclusive nature, but because it is unique in the sense that any two programs which list the language in the above manner are isomorphic.

Let $\mathscr{C}$ be a programming language, whose structure is described by a chain complex (C, Δ). Define two programs as being isomorphic, if one is mapped into the other by an isomorphism between two chain complexes, containing the two programs. Any two list-making programs for $\mathscr{C}$ of the desired kind are joined by an automorphism of (C, Δ). Hence, if isomorphic programs are regarded as representing the same algorithm, then it can be said that $\mathscr{C}$ has a unique list-making algorithm of the type described above. It will be proved that the input and instructions for any program, representing this algorithm, are complete sets of representatives of bases for the two quotient groups

$$(\text{kernel } \Delta)/\text{image } \Delta)$$

and

$$C/\text{kernel } \Delta),$$

respectively. Notice that the first of these is the *homology* group of the complex (C, Δ). Together they determine precisely the set of all programs which list without repetitions a basis for C and have minimum input. Selecting one program from this set, which lists the preferred basis $\mathscr{C}$, is our main problem. Algebraically it is the same as selecting a *standard basis* for a chain complex, a structure which is well-known in algebraic topology, as presented, for instance, in the textbook by Hilton and Wylie [3]. This problem will be fully described in Chapter I, and it will be solved for a variety of languages in the subsequent chapters. An understanding of finitely generated Abelian groups will be assumed. With this, there will be no difficulty in understanding the basic facts about chain complexes, as they come up.

The theory, presented in this book, is essentially a branch of combinatorial mathematics. The chain complexes, we use, are always finitely generated and can well be described as *combinatorial complexes*, since their main purpose is to display in a convenient notation the combinatorial structure of a programming language. This branch of combinatorics, which is concerned with listing, rather than counting the elements of a finite set, has been called 'algorithmic combinatorics' [1] or the study of 'combinatorial algorithms' [5]. All algorithms belong to this discipline, in the sense that an algorithm tells us how to write a program, which is essentially a list of instructions. Programming is a kind of list-making.

0.5. THE THEORY IN A NUTSHELL

It is the crux of our theory of algorithms that it approaches the problem of producing a desired output by finding a list of instructions which will produce the output. This means shifting our attention from the set $\mathscr{C}_1$, which contains the output, to the set $\mathscr{C}_2$ of instructions which are chosen as being suited to listing elements of $\mathscr{C}_1$.

If $\mathscr{C}_1$ were a family of sets, then $\mathscr{C}_2$ could be made up of instructions such as 'Form the union of A and B', provided that A, B, and $A \cup B$ are members of $\mathscr{C}_1$. A list of such instructions together with an input from $\mathscr{C}_1$ is all we need to generate each element of $\mathscr{C}_1$. Hence, the problem of listing every element of $\mathscr{C}_1$ becomes one of listing some elements of $\mathscr{C}_2$.

By way of solution to the problem of listing elements of $\mathscr{C}_1$ by means of instructions from $\mathscr{C}_2$, we look for a pair ($\mathscr{H}_1$, $\mathscr{D}_2$) of sets, defined as a *program*, where $\mathscr{H}_1$ is a subset of $\mathscr{C}_1$, called *input*, and $\mathscr{D}_2$ is a subset of $\mathscr{C}_2$, called the *instructions*, each of which produces a non-input element of $\mathscr{C}_1$. For this program to work there must exist a partial ordering of $\mathscr{D}_2$, such that each

instruction in it depends only on elements produced by previous
instructions or on elements of $\mathscr{H}_1$. The program is suggested by
the following diagram, showing the sets $\mathscr{C}_1$ and $\mathscr{C}_2$ as circles, each
of which has been broken into two parts:

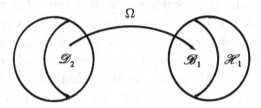

Ω is a one-to-one map of $\mathscr{D}_2$ onto $\mathscr{B}_1$, which carries each instruction to the element which it produces. Among all such programs
we confine our attention to those in which the input $\mathscr{H}_1$ has a
minimum of elements.

In general, there are many ways of choosing a program, and it is
our object to study the set of all of them. The next step in this
direction is to determine the elements of $(\mathscr{C}_2 - \mathscr{D}_2)$, those which
were not used in the program. This entails a new set $\mathscr{C}_3$ of instructions and a new program $(\mathscr{H}_2, \mathscr{D}_3)$ to list the elements of $\mathscr{C}_2$, not
already in $\mathscr{D}_2$. Therefore, if we pursue the objective of producing
as many as possible of the elements of our programming language,
we arrive at a graded language

$$\ldots, \mathscr{C}_3, \mathscr{C}_2, \mathscr{C}_1,$$

which separates into subsets as follows:

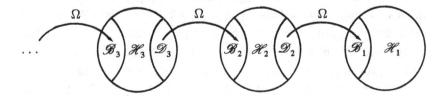

This diagram displays a program for listing all the elements of the language $\mathscr{C}$, where

$$\mathscr{C} = \mathscr{C}_1 \cup \mathscr{C}_2 \cup \mathscr{C}_3 \cup \ldots$$

In the same manner as with $\mathscr{C}$ let $\mathscr{H}$, $\mathscr{D}$, and $\mathscr{B}$ denote the union of the $\mathscr{H}$'s, $\mathscr{D}$'s, and $\mathscr{B}$'s; then $(\mathscr{H}, \mathscr{D})$ is a program which lists the elements of $\mathscr{C}$. $\mathscr{H}$ is the input, $\mathscr{D}$ the instructions, and $\mathscr{B}$ the set of elements which the instructions produce, Ω being the one-to-one function which connects each instruction with the element it produces. The diagram does not display the elements which each instruction depends on, nor how the elements of $\mathscr{D}$ are generated. These two properties of a program go together. Consider, for instance, the case when each instruction in $\mathscr{D}$ depends on exactly one element of $\mathscr{C}$, then the elements of $\mathscr{D}$ can be generated in the following way:

Assume we are in the midst of carrying out a program, and some, but not all, of the instructions in $\mathscr{D}$ have been generated, and only a partial list of those generated have been performed. Consequently, some, but not all, elements of $\mathscr{B}$ have been produced. Suppose b is the last element of $\mathscr{B}$ to have been listed, then at this point we take every instruction which depends on b and adjoin it to the partial list of $\mathscr{D}$. Eventually all of $\mathscr{D}$ is generated this way.

The structure of $\mathscr{C}$, displayed in the above diagram, captures the idea of programming language, as it will be understood in this book. It will be elaborated in various ways, but the essential form is there. Consequently, before going on to see how it can be made to display the elements which each instruction depends on and how other programs can be gotten from it, let us recognize the fact that it contains essentially all we need to know about a chain complex. The only feature of a chain complex, not displayed, is 'torsion', which could be included by simply allowing elements of

$\mathscr{B}$ to be integral multiples of elements of $\mathscr{C}$. This structure is what is known in homology theory as a *standard basis* for a chain complex and is known in the present theory as a *program*.

As an example of the connection with homology theory, notice the following formula, which is a direct consequence of the fact that Ω is a one-to-one correspondence: if $\mathscr{C}$ is finite, and if γ_π and η_π are the cardinalities of $\mathscr{C}_\pi$ and $\mathscr{H}_\pi$, respectively, then

$$\chi = \sum_\pi (-1)^\pi \gamma_\pi = \sum_\pi (-1)^\pi \eta_\pi.$$

This is known as the *Euler–Poincaré theorem*, when a chain complex is used to represent a polyhedron, and χ is known as its *Euler characteristic*.

Returning to the theory of algorithms, suppose an instruction d depends on a and produces b. Then, we can replace Ω by a new function Δ, such that

$$\Delta(d) = -a + b.$$

In order to include $(-a + b)$ in a language structure, similar to what has already been described, extend $\mathscr{C}$ to the free Abelian group which it generates, and extend Δ to a linear transformation. Then Δ is defined on the subgroup generated by $\mathscr{D}$, and we extend it to the entire group by giving it the value zero elsewhere. This is equivalent to giving Ω the value zero on all elements which were not instructions in the program described earlier. What has been done, therefore, is to replace the map Ω of $\mathscr{C}$ into itself by an endomorphism Δ of the free Abelian group generated by $\mathscr{C}$. This replaces a *language* by a *chain complex*.

The purpose of embedding a language in a chain complex is, not only for the convenience of including expressions such as $(-a + b)$, but more importantly to be able to define the set of all possible programs of a certain general kind. In particular, the set of all programs, which list (without repetitions and with minimum size input) all the elements of the programming language, can be defined and placed in a convenient algebraic setting. Properties of every program in the set can be deduced. However, to have defined a set of programs, which is complete in a given language, does not necessarily make it easy to find one of them. Nevertheless, the problem of finding a desired program has been formalized. Chapter I presents this formalization and some theory associated with it. The subsequent chapters are devoted to applications of the theory.

Chapter I

Programming

I.1. TO MAKE A LIST

Programming is viewed in this book as a mathematical subject, which is concerned with how to reduce a set, which is well-defined mathematically, to an explicit list of its elements. A *program* is a finite set of instructions, which, when carried out in a suitable order, will produce such a list. A formal definition of a program and a general approach to finding programs are the subject of this chapter.

Let us begin with what is to be defined later as an *elementary programming language*. It consists of two finite sets, $\mathscr{C}_1$ and $\mathscr{C}_2$, where $\mathscr{C}_1$ is the set which has to be listed, and $\mathscr{C}_2$ is the set of instructions which are allowed to be used in the process. Each instruction is a way of producing exactly one element of $\mathscr{C}_1$, by applying mathematical operations to other elements of $\mathscr{C}_1$. In this context a *program*, which lists without repetitions all the elements of $\mathscr{C}_1$, is a pair of sets: One is the *input* to the program, chosen from $\mathscr{C}_1$, and the other is the set *instructions* of the program, chosen from $\mathscr{C}_2$. The program is so constructed that every element of $\mathscr{C}_1$, not in the input, is produced by exactly one instruction, and, furthermore, there exists a linear ordering of the instructions such that each one depends only on elements produced by previous instructions or on elements of the input. These conditions are what is meant by 'listing the elements of $\mathscr{C}_1$ without repetition'.

When the linear ordering of the terms of a sequence is to be disregarded, they will be described as *entries* on a *list*. Accordingly, to *list* the elements of a finite set means to form any sequence whose terms belong to the set, such that every element of the set appears a designated number of times as a term of the sequence. Therefore, a list is characterized by, not only its set of entries, but also the number of times each entry appears. The most common way to list the elements of a set is *without repetitions*, in which case each element appears exactly once as an entry. The word *multiset* means the same as list, but it is inconvenient for our purposes, not being a verb.

Let us consider some examples of elementary languages, each denoted by $\mathscr{C}_1$ and $\mathscr{C}_2$, in which it is easy to see how a program might be written, using instructions from $\mathscr{C}_2$ to list all the elements of $\mathscr{C}_1$.

EXAMPLE I.1. Given μ letters, let $\mathscr{C}_1$ be the $\mu!$ sequences without repetitions of the μ letters, and let $\mathscr{C}_2$ be the $(\mu - 1)\mu!$ ways of choosing one such sequence and transposing an adjacent pair of letters in it. It is easy to see that with an input, consisting of one element of $\mathscr{C}_1$, there exists a sequence of $(\mu! - 1)$ elements of $\mathscr{C}_2$, as instructions, which will produce every other element of $\mathscr{C}_1$, in the following sense: Each instruction produces one element of $\mathscr{C}_1$ by applying a transposition to the input element or to an element produced by a previous instruction. The single input element together with the sequence of instructions constitute a *program* which lists without repetitions the elements of $\mathscr{C}_1$. A particularly interesting program would be one in which each instruction (except the first) applies a transposition to the element produced by the immediately preceding instruction.

EXAMPLE I.2. Given an integer μ, let $\mathscr{C}_1$ be all the partitions of μ (into positive integral summands), and let $\mathscr{C}_2$ be all the ways of choosing a partition of μ and adding two of its parts together. With an input, consisting of the partition of μ into ones, there exists a sequence of elements of $\mathscr{C}_2$, as instructions, which will produce, as output, every other element of $\mathscr{C}_1$ without repetitions.

EXAMPLE I.3. Given a set of μ elements, let $\mathscr{C}_1$ be its $(2^\mu - 1)$ non-void subsets, and let $\mathscr{C}_2$ be the set of all ways of forming a union of two elements of $\mathscr{C}_1$. With the list of all singleton subsets as input, there exists a sequence of elements of $\mathscr{C}_2$, as instructions, which will produce, as output, every non-singleton element of $\mathscr{C}_1$ without repetitions. This example is one in which each instruction applies a mathematical operation to more than one element of $\mathscr{C}_1$.

The above examples of languages involve only two kinds of expressions with which to write programs: Elements of $\mathscr{C}_1$, as input and output, and elements of $\mathscr{C}_2$, as instructions. But, just as there can be a recognized set $\mathscr{C}_2$ of ways in which some elements of $\mathscr{C}_1$ are derived from others, so there can be a recognized set $\mathscr{C}_3$ of ways in which some elements of $\mathscr{C}_2$ are derived from others. In general, a language will involve a sequence

$$\mathscr{C}_1, \mathscr{C}_2, \ldots, \mathscr{C}_\pi, \ldots, \mathscr{C}_\mu$$

of such sets, where $\mathscr{C}_\pi$ is said to be of *dimension π*.

In computer programming it is possible to recognize the separation of the language into dimensions in a comparable sense. That is, the dimension of an instruction is higher than that of the data upon which it operates. For example, dimensions 1, 2, and 3 could be as follows:

1. Integers up to a certain absolute value.
2. Machine language.
3. Assembly language.

I.2. ELEMENTARY PROGRAMMING LANGUAGES

A program cannot be written until a programming language has been chosen. The simplest kind of program, to be considered here, is that which lists the elements of a finite set $\mathscr{C}_1$ by means of instructions taken from a finite set $\mathscr{C}_2$, in which case the *language* of the program includes at the very least the set $\mathscr{C}_1 \cup \mathscr{C}_2$. Before looking at the formal definition of a programming language, let us consider the two kinds of elements which it must contain.

(i) *Input* and *output* elements: The existence of a program which lists all the elements of $\mathscr{C}_1$, using only instructions from $\mathscr{C}_2$, is assured by the fact that the program will include by definition a subset of $\mathscr{C}_1$, called the *input*, containing the elements which are not produced by the instructions in the program. This subset is required to be minimal, which means it is chosen so as to have the smallest number of elements consistent with the existence of the program. In the worst case, when $\mathscr{C}_2$ is void or contains no instructions capable of producing elements of $\mathscr{C}_1$, the input must contain every element of $\mathscr{C}_1$, and the program is trivial, in the sense that it contains no instructions.

(ii) *Instructions*: Each element d of $\mathscr{C}_2$ is a particular way in which a finite list $a_1, a_2, \ldots$ of elements of $\mathscr{C}_1$ may be operated upon mathematically, so as to produce a finite list $b_1, b_2, \ldots$. In elementary programming languages we confine ourselves to the special case, where the mathematical operation is a function of several variables. Suppose the instruction d is defined by a function f of ν variables, then it depends on a list

$$a_1, a_2, \ldots, a_\nu$$

of ν elements of $\mathscr{C}_1$ and produces a single element b of $\mathscr{C}_1$, determined by

$$b = f(a_1, a_2, \ldots, a_\nu).$$

The function is not necessarily symmetric, so that the instruction d must include, as one of its characteristics, the order in which the list is to be read.

Notice that we do not have the syntactical kind of instruction, known in computer programming as a 'transfer instruction', which governs the order in which instructions are to be performed. In fact, there is, in general, more than one linear order in which a program, as we shall define it, may be performed. Any ordering will be acceptable which places an instruction p ahead of an instruction q, whenever p produces one of the elements of $\mathscr{C}_1$ which q depends on. Consequently, the selection of a particular sequential ordering of instructions is a secondary problem, which is taken up after a program is completely determined. This problem will be dealt with in Section I.3.

DEFINITION I(1a). An *elementary programming language* is a pair $(\mathscr{C}_1, \mathscr{C}_2)$ of finite sets, such that each d in $\mathscr{C}_2$ is associated with a list of ν elements of $\mathscr{C}_1$, which d is said to *depend on*, where ν is a fixed non-negative integer, and with one element of $\mathscr{C}_1$, which d is said to *produce*.

DEFINITION I(1b). A *program in* $\mathscr{C}_2$, *which lists without repetitions all the elements of* $\mathscr{C}_1$, consists of a set of elements of $\mathscr{C}_1$, called the *input*, and set of elements of $\mathscr{C}_2$, called the *instructions*, which satisfy the following conditions: There exist linear orderings

$$h_1, h_2, \ldots h_\eta \quad \text{and} \quad d_1, d_2, \ldots, d_\beta$$

of the input and instructions, respectively, such that, if b_θ denotes the element of $\mathscr{C}_1$ produced by d_θ, then
(i) d_θ depends only on entries in the list

$$h_1, h_2, \ldots, h_\eta, b_1, b_2, \ldots, b_{\theta-1}.$$

(ii) Every element of $\mathscr{C}_1$ appears once in the list

$$h_1, h_2, \ldots, h_\eta, b_1, b_2, \ldots, b_\beta.$$

(iii) The input is minimal; that is, η has the smallest value for which the program, as otherwise defined, exists.

These definitions cover a special situation, associated with the non-negative integer ν. In the general case, considered later, different elements of a language may depend on different numbers of elements and produce different numbers of elements. In such a language it may still be possible to construct an *elementary program*, that is, one in which each instruction depends on ν elements and produces one element. Let us consider elementary languages and programs for small values of ν.

(i) $\nu = 0$. In this case each element of $\mathscr{C}_2$ produces one element of $\mathscr{C}_1$ and depends on none. It is characterized by a function

$$\Delta: \mathscr{C}_2 \to \mathscr{C}_1.$$

A program in this language, which lists the elements of $\mathscr{C}_1$, is defined by any restriction of the function Δ, which is one-to-one and has the same image as Δ. The *input* consists of all elements not in the image of Δ; the *instructions* are the elements to which Δ is restricted; and the *output* is the image of Δ. This simple kind of program will appear later, only as a step in developing more useful programs.

(ii) $\nu = 1$. This is called the *graphical* case, and it is discussed at length in Section I.4. Such a language is characterized by a function of $\mathscr{C}_2$ into $\mathscr{C}_1 \times \mathscr{C}_1$, and it can be regarded as an oriented graph, whose vertices and edges are the elements of $\mathscr{C}_1$ and $\mathscr{C}_2$, respectively. Each edge leads from the single vertex which it depends on to the single vertex which it produces. A program in this language is an oriented subgraph, in which no edges are oriented so as to lead to an input vertex, and exactly one edge is oriented so as to lead to each output vertex.

(iii) $\nu = 2$. This case may be illustrated by a simple example. Let $\mathscr{C}_1$ be the divisors greater than 1 of 30; let $\mathscr{C}_2$ consist of every multiplication $j \times k$, such that j, k, and their product jk belong to $\mathscr{C}_1$. Then a program, for listing the elements of $\mathscr{C}_1$, has *input*

$$2, 3, 5$$

and *instructions*

$$2 \times 3, 2 \times 5, 3 \times 5, 2 \times 15.$$

The *output* is

$$6, 10, 15, 30.$$

Each instruction $j \times k$ depends on two elements j and k of $\mathscr{C}_1$, and produces one element jk.

I.3. TO SPECIFY A PROGRAM INDUCTIVELY

In Definition I.1 it was taken into account that there is not always a unique linear order in which the instructions of a program must be carried out. The set of instructions may be separable into parts, which are independent of each other and able to be performed simultaneously or in any order. If we wish to be completely explicit about the way in which a program is to be performed, it is always possible in principle to do so, by choosing one of the linear orderings which satisfies definition I(1b). To find such a linear ordering, and to express it suitably constitute the third step in a total process which may be described as 'finding an algorithm'. The steps are described as follows:

(i) Choosing a language $\mathscr{C}$.

(ii) Finding a program in $\mathscr{C}$.

(iii) Specifying the program.

One way to specify a program would be to write down the input and instructions of the program in the form of two sequences,

$$(h_1, h_2, \ldots, h_\eta) \text{ and } (d_1, d_2, \ldots d_\beta),$$

so as to satisfy Definition I(1b). It is comparable to writing a computer program without the use of any transfer instructions, so that the computer performs every coded instruction exactly once and in the order given. Instead of this let us specify a program inductively, and thus introduce into our total concept of an algorithm what would be described in computer programming as *loops*. What it amounts to mathematically is definition by induction of the sequence

$$(d_1, d_2, \ldots, d_\beta)$$

of instructions.

We start with a given program satisfying Definition I(1b), which lists without repetitions the elements of $\mathscr{C}_1$ by means of instructions from $\mathscr{C}_2$. Two minor assumptions are needed: That the input is not empty and that $\nu = 1$. It will be shown later that any program can be made to satisfy the second assumption. Actually, the method, which is about to be described for specifying a program inductively can be generalized easily to cover the cases when $\nu > 1$.

There exists a partition of the set of instructions of the given program which is defined by putting any two instructions in the same part when they depend on the same element of $\mathscr{C}_1$. It is clear from Definition I(1b) that it makes no difference how the elements within a single part are linearly ordered. For any $a \in \mathscr{C}_1$ let $\mathscr{L}(a)$

be a list which contains without repetitions all the instructions which depend on a, and assume that we have the ability to produce such a list for all a. This is a significant assumption, but it is much less than the ability to place all the instructions in a suitable sequence, since the list $\mathscr{L}(a)$ is allowed to have its entries in any order and is much shorter, in general, than an entire list of the elements of $\mathscr{C}_1$.

If $\mathscr{L}$ is given, rather than an explicit list of instructions, then the program is completely specified by the following mathematical induction: There exists a first entry h_1 in the list of elements which the program is expected to produce, because an input list

$$h_1, h_2, \ldots, h_\eta$$

is given and is not empty. Next, make the inductive assumption that a partial list

$$h_1, h_2, \ldots, h_\eta, b_1, b_2, \ldots, b_\phi$$

is known, where $b_1, b_2, \ldots, b_\phi$ are the elements of $\mathscr{C}_1$ which have been produced by the instructions $d_1, d_2, \ldots, d_\phi$, respectively. In particular, d_ϕ produces b_ϕ and depends on some term which precedes b_ϕ in the partial sequence. Therefore, the immediate successor of this term is still in the partial sequence, and let us denote it by a. In other words, a is the term which follows that which d_ϕ depends on. There is a list $\mathscr{L}(a)$, whose entries we can denote by

$$d_{\phi+1}, d_{\phi+2}, \ldots,$$

and the list of entries, which these produce in corresponding order, are

$$b_{\phi+1}, b_{\phi+2}, \ldots.$$

This is adjoined to the partial list which was given, which completes the induction, unless $\mathscr{L}(a)$ is void, in which case $\mathscr{L}$ is applied to the successor of a. If a does not have a successor, then the list of elements of $\mathscr{C}_1$, which was to have been formed, is complete.

In computer programming a program can be specified by a flow-chart. If all instructions are given explicitly, the flow chart will be a sequence of boxes, each containing a description of a computational step, with an arrow leading from one box to the next, showing the order in which the steps have to be taken. If the program is specified inductively the arrows will form a closed loop. Figure 1 (also flowchart 1 of the appendix) specifies a program in $\mathscr{C}_2$ which lists the elements of $\mathscr{C}_1$, where $\mathscr{H}_1$ and the list-making operator $\mathscr{L}$ are assumed to be known.

EXAMPLE I.4. A program for listing without repetitions every permutation of a given n-termed sequence $\mathscr{S}$ can be written in the following language:

$\mathscr{C}_1$ is the set of all permutations of the sequence $\mathscr{S}$. (This is a generalization of Example I.1, because $\mathscr{S}$ is allowed to contain repeated elements, which means that the number of permutations of $\mathscr{S}$ is

$$n!/n_1!n_2!\ldots,$$

where $n_1, n_2, \ldots$ are the numbers of times that the various terms appear in $\mathscr{S}$.)

$\mathscr{C}_2$ is the set of all ways of choosing an element of $\mathscr{C}_1$ and transposing an adjacent pair of distinct terms in it.

Now, a program in this language for listing without repetitions all elements of $\mathscr{C}_1$ is completely determined, if the following input and the following $\mathscr{L}(x)$ are explicitly given:

The input is one element of $\mathscr{C}_1$, which is constructed by assigning a linear ordering to the set of terms in $\mathscr{S}$ and by taking that element of $\mathscr{C}_1$ in which the terms are in non-decreasing order.

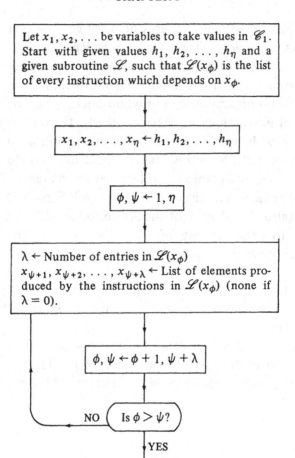

Fig. 1. This is the flowchart of a program, specified in the manner of this section. The variables $x_1, x_2, \ldots$ take on values in $\mathscr{C}_1$, and the variables λ, ϕ, ψ taken on integer values. $v \leftarrow k$ means that a variable expression v takes on the value of a known expression k.

$\mathscr{L}(x)$ is a list without repetitions of each way of transposing an adjacent pair of terms in x, such that after transposition it will become the first decreasing pair of terms in x.

A proof is required that this is a correct program. For the time being, let us verify that it works in the case of the sequence:

$$\mathscr{S} = (a, b, c, c).$$

If the set of terms is given the alphabetical ordering, then the input has the following single entry:

$$h_1 = a, b, c, c.$$

$\mathscr{L}(h_1)$ is the following list with 2 entries:

$$d_1 = b \times a, c, c,$$
$$d_2 = a, c \times b, c.$$

Here an element of $\mathscr{C}_2$ is denoted by putting a cross in place of a comma between the transposed terms. This list is known. The instructions d_1 and d_2 produce b_1 and b_2:

$$b_1 = b, a, c, c$$
$$b_2 = a, c, b, c.$$

Now we continue the procedure:

$$\mathscr{L}(b_1) = d_3 = b, c \times a, c$$
$$b_3 = b, c, a, c$$
$$\mathscr{L}(b_2) = \begin{cases} d_4 = c \times a, b, c \\ d_5 = a, c, c \times b \end{cases}$$

$$b_4 = c, a, b, c$$

$$b_5 = a, c, c, b$$

$$\mathscr{L}(b_3) = \begin{cases} d_6 = c \times b, a, c \\ d_7 = b, c, c \times a \end{cases}$$

$$b_6 = c, b, a, c$$

$$b_7 = b, c, c, a$$

$\mathscr{L}(b_4) = 0$ (The list with no entries.)

$$\mathscr{L}(b_5) = d_8 = c \times a, c, b$$

$$b_8 = c, a, c, b$$

$$\mathscr{L}(b_6) = 0$$

$$\mathscr{L}(b_7) = d_9 = c \times b, c, a$$

$$b_9 = c, b, c, a$$

$$\mathscr{L}(b_8) = d_{10} = c, c \times a, b$$

$$b_{10} = c, c, a, b$$

$$\mathscr{L}(b_9) = d_{11} = c, c \times b, a$$

$$b_{11} = c, c, b, a$$

$$\mathscr{L}(b_{10}) = 0$$

$$\mathscr{L}(b_{11}) = 0.$$

We can see that the input h_1 together with the output $b_1, b_2, \ldots,$ b_{11} form the desired list of the 12 permutations of the given sequence.

EXAMPLE I.5. This depends on Example 1.4, where a program was demonstrated, which listed the permutations $\mathscr{C}_1$ of the sequence (a, b, c, c). The program also generated its own instructions which constituted a partial list $\mathscr{D}_2$ of elements of $\mathscr{C}_2$. Now let us find a program for listing the elements of $\mathscr{C}_2$, by means of instructions from a new set $\mathscr{C}_3$. Let us use as input the known partial list $\mathscr{D}_2$, as well as the smallest number of additional entries needed for the program to exist. The programming language is chosen as follows:

$\mathscr{C}_2$ is the set of all ways of taking one of the 12 permutations of (a, b, c, c) and transposing an increasing pair of adjacent letters in it. We do not bother with the decreasing pairs, since the elements, so defined, are in one-to-one correspondence with the chosen ones. The elements of $\mathscr{C}_2$ are written, as before, by putting a cross in place of a comma between the transposed pair.

$\mathscr{C}_3$ is the set of all ways of transposing an adjacent pair of terms in an element of $\mathscr{C}_2$, where $j \times k$ is regarded as a single term. (The only possible values of $j \times k$ are $b \times a, c \times a$, and $c \times b$.)

In the language $(\mathscr{C}_2, \mathscr{C}_3)$ a program for listing the elements of $\mathscr{C}_2$ requires no input other than

$$d_1, d_2, \ldots, d_{11}$$

taken from Example I.4. Relabel this input as

$$h_1, h_2, \ldots, h_{11}.$$

With this input a program of the desired kind will be completely determined if the following lists are explicitly given for each $x \in \mathscr{C}_2$:

$\mathscr{L}(x)$ is a list (without repetitions) of all ways of transposing an adjacent pair of terms in x of the form $j \times k, l$ or j, l, provided that after transposition l, j is the first decreasing pair of letters in x. (This is an obvious way of generalizing the $\mathscr{L}(x)$ in Example I.4.) The entries in $\mathscr{L}(x)$ belong to $\mathscr{C}_3$ and will be denoted as follows:

If $j \times k$, l is being transposed, replace it in x by $l \times j \times k$, and, if j, l is being transposed, replace it in x by $l \times j$.

Now we are prepared to demonstrate the program explicitly:

$$h_1 = b \times a, c, c$$

$$h_2 = a, c \times b, c$$

$$h_3 = b, c \times a, c$$

$$h_4 = c \times a, b, c$$

$$h_5 = a, c, c \times b$$

$$h_6 = c \times b, a, c$$

$$h_7 = b, c, c \times a$$

$$h_8 = c \times a, c, b$$

$$h_9 = c \times b, c, a$$

$$h_{10} = c, c \times a, b$$

$$h_{11} = c, c \times b, a.$$

In what follows we omit $\mathscr{L}(x)$ if it is empty:

$$\mathscr{L}(h_1) = d_1 = c \times b \times a, c$$

$$b_1 = c, b \times a, c$$

$$\mathscr{L}(h_5) = d_2 = c \times a, c \times b$$

$$b_2 = c, a, c \times b$$

$$\mathscr{L}(h_7) = d_3 = c \times b, c \times a$$

$$b_3 = c, b, c \times a$$

$$\mathcal{L}(b_1) = d_4 = c, c \times b \times a$$

$$b_4 = c, c, b \times a.$$

This shows that $\mathscr{C}_2$ has 15 elements, and this can be verified easily by looking at the list of 12 elements in $\mathscr{C}_1$ and counting the number of increasing pairs of adjacent letters which appear among them.

These examples have shown how, once a programming language is chosen, programs may be written. Now let us return to a consideration of the language itself.

I.4. GRAPHICAL PROGRAMMING LANGUAGES

A *graphical program* is one in which each instruction depends on exactly one element and produces exactly one element. This rules out, for instance, a multiplication instruction, which depends on two factors, unless we can define one factor as being 'fixed', so that the instruction depends only on the other. In this section let us confine our attention to a programming language ($\mathscr{C}_1$, $\mathscr{C}_2$), in which each element d of $\mathscr{C}_2$ depends on exactly one element a of $\mathscr{C}_1$ and produces exactly one element b of $\mathscr{C}_1$. This language can be characterized by an oriented graph, having $\mathscr{C}_1$ as its set of vertices and $\mathscr{C}_2$ as its set of edges, such that d is the edge leading from a to b. We shall arrive at the full range of programming languages to be used in this book, by extending this type of language, so as to make programming easier by having more instructions available. As a first step in this direction let us extend ($\mathscr{C}_1$, $\mathscr{C}_2$), so that it may be characterized by an unoriented graph.

An elementary programming language ($\mathscr{C}_1$, $\mathscr{C}_2$) is *graphical* (in the unoriented sense), if $\nu = 1$, and, if with each element d of $\mathscr{C}_2$, which depends on a and produces b, there exists another element

$-d$ of $\mathscr{C}_2$, which depends on b and produces a. The graph $\mathscr{G}$ which represents this language has one edge for both d and $-d$, and a program in the language is a subgraph of $\mathscr{G}$, whose edges are the instructions and whose vertices are the input and output.

THEOREM I.1. *Let $\mathscr{G}$ be the graph defined by a graphical program-ming language ($\mathscr{C}_1$, $\mathscr{C}_2$), and let $\mathscr{F}$ be the subgraph defined by any program in $\mathscr{C}_2$, which lists without repetitions the elements of $\mathscr{C}_1$. Then, $\mathscr{F}$ consists of a maximal tree in each connected com-ponent of $\mathscr{G}$, and the input corresponds to one vertex of each tree.*

Proof. Let γ be the number of vertices in $\mathscr{G}$, and let η be the number of connected components of $\mathscr{G}$. If we started with any maximal tree in each component of $\mathscr{G}$ and one arbitrarily chosen vertex from each tree, it is easy to show the existence of a pro-gram, having the chosen vertices as input and the edges of the trees as instructions. This program would have η input elements and $\gamma - \eta$ instructions. No program can have less than η input elements, because no path in $\mathscr{G}$ could pass from one component to another, and no program can have more than η input elements by part (iii) of definition I(1b).

Therefore, any program in $\mathscr{C}_2$, which lists the elements of $\mathscr{C}_1$ has η input elements and $\gamma - \eta$ instructions. The corresponding graph $\mathscr{F}$ has at least η connected components, because it spans the vertices of $\mathscr{G}$. If it had more than η components, there would be one with no input vertex, only output vertices. However, each output vertex is connected by a path in $\mathscr{F}$ to an input vertex. Hence $\mathscr{F}$ has η components, γ vertices, and $\gamma - \eta$ edges. Under these circumstances $\mathscr{F}$ has no cycles or closed paths, and must consist of trees. This proves the theorem.

A graphical program in $\mathscr{C}_2$, which lists without repetitions the elements of $\mathscr{C}_1$, never includes both an instruction d and its inverse

$-d$. Therefore, the graph $\mathscr{F}$ of the program has an orientation defined by its instructions. If an instruction depends on a and produces b, then the corresponding edge in $\mathscr{F}$ leads from vertex a to vertex b. To specify the program by the method introduced in Section 1.3 is equivalent to the following: Each input vertex of $\mathscr{F}$ is given, and a function $\mathscr{L}$ is given which assigns to each vertex a the list $\mathscr{L}(a)$ of all edges which lead from a. This determines $\mathscr{F}$ and specifies the program.

If the program is the kind in which each instruction produces the element, which the very next instruction depends on, then for each a there is exactly one entry in $\mathscr{L}(a)$, and $\mathscr{F}$ consists of a *Hamiltonian* path in each component of $\mathscr{G}$.

The transition from graphical to non-graphical programming languages is very easily made, when the languages are embedded in chain complexes. In the graphical case a language $(\mathscr{C}_1, \mathscr{C}_2)$ is a basis for the simplicial complex associated with the graph $\mathscr{G}$ of the language. The *chain* groups C_1 and C_2 of the complex are the free Abelian groups, generated by $\mathscr{C}_1$ and $\mathscr{C}_2$, respectively, and the *boundary homomorphism*

$$\Delta: C_2 \to C_1$$

is determined by

$$\Delta(d) = -a + b,$$

where d is an element of $\mathscr{C}_2$, which depends on a and produces b. This is the standard way of describing the simplicial complex associated with $\mathscr{G}$, except that the subscripts are conventionally one less.

I.5. CHAIN COMPLEXES

From now on every programming language will be embedded in a *chain complex*, which is defined as a sequence

$$0 \to C_\mu \to C_{\mu-1} \to \ldots \to C_2 \to C_1 \to 0$$

of homomorphisms (shown by arrows) between finitely generated free Abelian groups, called *chain groups*, such that the first and last homomorphisms are trivial, and the composition of two successive ones is trivial. In other words, if for any π the homomorphism of C_π is denoted by

$$\Delta_\pi : C_\pi \to C_{\pi-1},$$

then for any $x \in C_{\pi+1}$

$$\Delta_\pi(\Delta_{\pi+1}(x)) = 0.$$

It is assumed that for $\pi < 1$ and $\pi > \mu$

$$C_\pi = 0.$$

Each Δ_π is called a *boundary function*.

The following groups are defined for each π, which is called the *dimension* of the group:

B_π is the image of $\Delta_{\pi+1}$ and is called the group of *boundaries* in C_π.

$$B_\pi = \{\Delta_{\pi+1}(x) : x \in C_{\pi+1}\}.$$

Z_π is the kernel of Δ_π and is called the group of *cycles* in C_π.

$$Z_\pi = \{x : \Delta_\pi(x) = 0\}.$$

H_π is the quotient group Z_π/B_π and is called the πth *homology group*.

$$H_\pi = \{(z + B_\pi) : z \in Z_\pi\}.$$

D_π is the quotient group C_π/Z_π. It is isomorphic to $B_{\pi-1}$.

$$D_\pi = \{(x + Z_\rho) : x \in C_\pi\}.$$

The above complex can be described without reference to dimensions as the pair (C, Δ), where C is the direct sum

$$C = C_1 + C_2 + \ldots + C_\pi + \ldots + C_\mu,$$

and Δ is the unique endomorphism of C whose restriction to C_π is Δ_π. The preferred basis for C is

$$\mathscr{C} = \mathscr{C}_1 \cup \mathscr{C}_2 \cup \ldots \cup \mathscr{C}_\mu.$$

The important groups associated with (C, Δ) are as follows:

$$Z = \text{kernel of } \Delta,$$
$$B = \text{image of } \Delta,$$
$$H = Z/B,$$
$$D = C/Z.$$

To illustrate the significance of each of these groups, let us return to the simplicial complex of the last section, which is generated by an elementary graphical language $(\mathscr{C}_1, \mathscr{C}_2)$. Every chain group is trivial except C_1 and C_2, which are generated by $\mathscr{C}_1$ and $\mathscr{C}_2$, respectively, and every boundary function is trivial except Δ_2, which was denoted previously by Δ. Therefore, the complex is of the form

$$0 \to C_2 \to C_1 \to 0.$$

The next theorem concerns the two quotient groups H_1 and D_2

for this complex. It shows essentially that in any program in $\mathcal{C}_2$, which lists without repetitions the elements of $\mathcal{C}_1$, the input represents a basis for H_1, and the instructions represent a basis for D_2. This fact, which will apply to any language, is now shown in the graphical case.

THEOREM I.2. *Let*

$$0 \rightarrow C_2 \rightarrow C_1 \rightarrow 0$$

be the simplicial complex of an elementary graphical language $(\mathcal{C}_1, \mathcal{C}_2)$, *and let*

$$h_1, h_2, \ldots, h_\eta \text{ and } d_1, d_2, \ldots, d_\beta$$

be the input and instructions, respectively, of any program in $\mathcal{C}_2$, *which lists without repetitions the elements of* $\mathcal{C}_1$. *Then, the cosets*

$$(h_1 + B_1), (h_2 + B_1), \ldots, (h_\eta + B_1)$$

form a basis of H_1, *and the cosets*

$$(d_1 + Z_2), (d_2 + Z_2), \ldots, (d_\beta + Z_2)$$

form a basis of D_2.

   Proof. Let $b_1, b_2, \ldots, b_\beta$ be the elements produced by the instructions of the given program. Then, according to Definition I(1b),
$$h_1, h_2, \ldots h_\eta, b_1, b_2, \ldots, b_\beta$$

lists the elements of $\mathcal{C}_1$. Also, it is a basis for C_1. To prove that the cosets

$$(h_1 + B_1), (h_2 + B_1), \ldots, (h_\eta + B_1)$$

are a basis for H_1, we show that they span H_1 and are linearly independent.

The cosets span H_1, if for each x in the given basis for C_1, the coset $(x + B_1)$ is a linear combination of the given cosets. In fact, we can show that $(x + B_1)$ equals one of the cosets, which is obvious, when x is one of the h's. Therefore, it remains to show that for any θ there is ϕ such that

$$(b_\theta + B_1) = (h_\phi + B_1).$$

That is,

$$(b_\theta - h_\phi) \in B_1.$$

For any ϕ the instruction which produces b_ϕ is denoted by d_ϕ. Consider the series of instructions

$$d_\theta + \ldots + d_\phi + \ldots$$

such that:

(i) if d_ϕ depends on an output element b_ψ, then d_ϕ is followed by d_ψ.

(ii) if d_ϕ depends on an input element h_ψ, then d_ϕ is the last term of the series.

It follows that, if Δ is the boundary function,

$$\Delta(d_\theta + \ldots + d_\phi + \ldots) = -h_\psi + b_\theta.$$

Hence

$$(b_\theta - h_\phi) \in B_1.$$

To prove that the cosets are independent, we can use the well known fact that the lowest homology group of a graph is a free Abelian group, whose rank is the number of connected components of the graph. In Theorem I.1 we showed that the complex in this theorem belongs to a graph with η connected components. Hence, η is the rank of H_1, and the η cosets, which span H_1, must be independent.

For all θ,

$$\Delta(d_\theta) = -a_\theta + b_\theta,$$

where

$$a_\theta \in \{h_1, \ldots, h_\eta, b_1, \ldots, b_{\theta-1}\}.$$

It follows from this that the basis for C_1, given by the program, can be changed to a new basis

$$h_1, \ldots, h_\eta, \Delta(d_1), \ldots, \Delta(d_\beta).$$

It has been seen that no non-trivial linear combinations of the h's are in B_1. Therefore,

$$\Delta(d_1), \ldots, \Delta(d_\beta)$$

is a basis for B_1. By the isomorphism between D_2 and B_1 we conclude that

$$(d_1 + Z_2), (d_2 + Z_2), \ldots, (d_\beta + Z_2)$$

is a basis for D_2. This completes the proof.

I.6. PROGRAMMING COMPLEXES

A programming language $\mathscr{C}$ will now be extended to an infinite counterpart C, called a *programming complex*, which is uniquely determined by $\mathscr{C}$. The purpose of this extension is to make the programming easier, by defining a larger set of elements which may be used as instructions.

Let us start with the elementary graphical programming language, which has been examined in the last two sections. It consists of a pair ($\mathscr{C}_1$, $\mathscr{C}_2$) of finite sets, where each element d of $\mathscr{C}_2$ determines an ordered pair (a, b) of elements of $\mathscr{C}_1$, where d is said to *depend on a* and *produce b*. In Section I.6.1, $\mathscr{C}_1$ and $\mathscr{C}_2$ will be extended to include formal sums and differences of their elements and, in Section I.6.2, to include higher dimensional analogs, $\mathscr{C}_3$, $\mathscr{C}_4$, ...

I.6.1 Formal Sums

($\mathscr{C}_1$, $\mathscr{C}_2$) is redefined, to allow each element d of $\mathscr{C}_2$ to depend on a finite list $a_1, a_2, \ldots$ of elements of $\mathscr{C}_1$ and to produce a finite list $b_1, b_2, \ldots$ of elements of $\mathscr{C}_1$. This is done by expressing the lists as formal sums

$$a_1 + a_2 + \ldots \quad \text{and} \quad b_1 + b_2 + \ldots$$

and describing the new structure by the homomorphism

$$\Delta : C_2 \to C_1,$$

where C_1 and C_2 are the free Abelian groups generated by $\mathscr{C}_1$ and $\mathscr{C}_2$, respectively, and Δ is the unique homomorphism defined by

$$\Delta(d) = -(a_1 + a_2 + \ldots) + (b_1 + b_2 + \ldots).$$

A new kind of joint instruction is available in C_2, consisting of a finite list $d_1, d_2, \ldots$ of elements of $\mathscr{C}_2$. It is written as a formal sum
$$d_1 + d_2 + \ldots ,$$

and by virtue of the linearity of Δ, it depends on all the elements which its terms depend on, and produces all the elements which its terms produce.

DEFINITION I(2a). A *short programming complex* is a chain complex

$$0 \to C_2 \to C_1 \to 0,$$

in which C_1 and C_2 are finitely generated free Abelian groups. It is denoted by its only non-trivial homomorphism,

$$\Delta : C_2 \to C_1,$$

called the *boundary function*. If the first homology group H_1 is a free group, where

$$H_1 = C_1/\text{im } \Delta,$$

then the complex is said to be *free*.

For the rest of this section we shall confine our attention to free complexes. From the programming viewpoint, what it means for the complex to be free, is that, if C_2 contains an instruction which produces a multiple of an element of C_1, then it also contains an instruction which produces the element itself. Any bases $\mathscr{C}_1$ and $\mathscr{C}_2$ for C_1 and C_2, respectively, constitute a *programming language* in the complex. The language is *free*, if the complex is free.

DEFINITION I(2b). In a short free programming complex

$$\Delta : C_2 \to C_1$$

a program in C_2, *which lists a basis for* C_1, *consists of an input*

$$h_1, h_2, \ldots, h_\eta,$$

selected from C_1, and *instructions*

$$d_1, d_2, \ldots, d_\beta,$$

selected from C_2, such that:
(i) for $\theta = 1, 2, \ldots, \beta$

$$\Delta(d_\theta) = -a_\theta + b_\theta,$$

where b_θ is an element of C_1 and a_θ is in the group generated by

$$h_1, h_2, \ldots, h_\eta, b_1, b_2, \ldots, b_{\theta-1}.$$

(ii) the list

$$h_1, h_2, \ldots, h_\eta, b_1, b_2, \ldots, b_\beta$$

is a basis for C_1.
(iii) η is the smallest value for which the above conditions can be satisfied.

THEOREM I.3. *Let* $\Delta : C_2 \to C_1$ *be a short free programming complex. Then,*

$$h_1, h_2, \ldots, h_\eta$$
and
$$d_1, d_2, \ldots, d_\beta$$

are the input and instructions of a program in C_2, *which lists a basis for* C_1, *if, and only if,*

$$(h_1 + B_1), (h_2 + B_1), \ldots, (h_\eta + B_1)$$
and

$$(d_1 + Z_2), (d_2 + Z_2), \ldots, (d_\beta + Z_2)$$

are bases for H_1 and D_2, respectively.

In the above theorem the chain complex is given. Therefore, the groups B_1, Z_2, H_1, and D_2 are completely determined by the definitions of Section I.5. It follows that the constants η and β are attributes of the complex, so that every program in C_2, which lists a basis for C_1, has exactly η input elements and β instructions.

Proof. Let us begin with given bases for H_1 and D_2. By the homomorphism theorem for groups D_2 is isomorphic to B_1 under the map which takes $(d_\theta + Z_2)$ to b_θ, where

$$b_\theta = \Delta(d_\theta).$$

Thus, condition (i) of Definition I(2b) is satisfied trivially, and condition (ii) follows from the fact that under this isomorphism the b_θ's form a basis for B_1, and the h_θ's extend it to a basis for C_1. Condition (iii) follows from the fact that any program has its instructions in one-to-one correspondence with a linearly independent set of boundaries, so that no program can have more than β instructions. Since $\eta + \beta$ is the rank of C_1, no program can have less than η input elements.

Now consider the converse. H_1 is a free group by assumption, and D_2 is a free group, because it is isomorphic to a subgroup B_1 of a free group C_1. Therefore, there exist bases for H_1 and D_2, and by the first half of the proof there exists a program with η input elements and β instructions, where η and β are the ranks of H_1 and D_2, respectively. Consequently, any program in C_2, which lists a basis for C_1, has η input elements

$$h_1, h_2, \ldots, h_\eta$$

and β instructions

$$d_1, d_2, \ldots, d_\beta.$$

It follows from Definition I(2b) that

$$h_1, h_2, \ldots, h_\eta, \Delta(d_1), \Delta(d_2), \ldots, \Delta(d_\beta)$$

is a basis for C_1, because every b_θ is expressible as a linear combination of these. The elements of the form $\Delta(d_\theta)$ constitute a maximal linearly independent subset of B_1, so that no non-trivial linear combination of the h_θ's is in B_1. But every element of C_1 differs from some linear combination of the h_θ's by an element of B_1. It follows that

$$(h_1 + B_1), (h_2 + B_1), \ldots, (h_\eta + B_1)$$

is a basis for H_1, and also that

$$\Delta(d_1), \Delta(d_2), \ldots, \Delta(d_\beta)$$

is a basis for B_1. By the isomorphism, mentioned above, between B_1 and D_2, we conclude that

$$(d_1 + Z_2), (d_2 + Z_2), \ldots, (d_\beta + Z_2)$$

is a basis for D_2. This completes the proof.

A program under Definition I(2b) is *graphical*, if for all θ the term a_θ is, not merely linearly dependent on

$$h_1, h_2, \ldots, h_\eta, b_1, b_2, \ldots, b_{\theta-1},$$

but actually appears on this list.

THEOREM I.4. *Let $C_2 \to C_1$ be a short programming complex. If H_1 is a non-trivial free group, then there exists a graphical program in C_2, which lists a basis for C_1.*

Proof. Since H_1 is free, there exists a basis for C_1 of the form

$$h_1, h_2, \ldots, h_\eta, \Delta(d_1), \Delta(d_2), \ldots, \Delta(d_\beta),$$

in which the Δ terms are a basis for B_1.

Since H_1 is non-trivial, $\eta \geqslant 1$ and, therefore, h_1 exists. Let

$$b_\theta = \Delta(d_\theta) + h_1$$

for $\theta = 1, 2, \ldots, \beta$.

It follows that

$$h_1, h_2, \ldots, h_\eta \quad \text{and} \quad d_1, d_2, \ldots, d_\beta$$

are the input and instructions of a graphical program in C_2 which generates a basis for C_1.

Thus the theorem has been proved by exhibiting a trivial example, where the graph of the program is such that every edge radiates from the single vertex h_1, and the remaining input vertices are isolated points.

I.6.2. Higher Dimensions

So far, a programming language has been separated into two dimensions, the first being used for data and the second for instructions. A language ($\mathscr{C}_1$, $\mathscr{C}_2$) and its extension to a complex

$$0 \to C_2 \to C_1 \to 0$$

have been defined, so as to allow a program to be written which will list data, but not instructions. Our object now is to define a language $\mathscr{C}$ and its extension to a complex (C, Δ), in which a program can be written to list any element of $\mathscr{C}$, which means that it is possible for instructions to be regarded as data, for the purpose of listing them. Computer languages are of this nature, since instructions can always be generated, stored in memory, and treated as data. We shall accomplish this by replacing the homomorphism Δ of C_2 into C_1, where

$$C_1 \cap C_2 = 0,$$

by an endomorphism Δ, which carries a single group C into itself. Thus, in language $\mathscr{C}$, which is a basis for C, the distinction between data and instructions is removed, and is only recognized in particular programs, according to how such elements are being used.

The complex (C, Δ) is constructed by starting with a short complex $C_2 \to C_1$ and introducing a new short complex $C_3 \to C_2$, in which there exists a program which lists a basis for C_2. The process is done repeatedly until we arrive at a sequence

$$0 \to C_\mu \to C_{\mu-1} \to \ldots \to C_\pi \to C_{\pi-1} \to \ldots \to C_1 \to 0$$

of homomorphisms, originating with a trivial one $0 \to C_\mu$. C_π is called the *chain group of dimension* π. We require that no two chain groups have any elements except 0 in common, and we define C as the direct sum of all the chain groups.

$$C = C_1 + C_2 + \ldots + C_\mu.$$

Each homomorphism is denoted by Δ, which extends uniquely to an endomorphism of C, also written as Δ.

For $\pi > 1$, by making each new short complex as small as possible, or, at least, not unnecessarily large, we improve our chances of arriving at a trivial homomorphism after a finite number of steps. This is required, since C is finitely generated. Each pair, $(C_{\pi-1}, C_{\pi})$ and $(C_{\pi}, C_{\pi+1})$, of languages, as described above, involves a redundancy, because a program in C_{π}, which lists a basis for $C_{\pi-1}$, also determines a list of instructions in C_{π}, which do not then need to be listed again as part of the next program in $C_{\pi+1}$. It is clear from Theorem I.3 that the program in $C_{\pi+1}$ only needs to list a basis for Z_{π}, since this basis together with the previously determined instructions from C_{π} will constitute a basis for C_{π}.

In other words, to construct a program in C, which lists a basis for C, we use a sequence of programs, one in each C_{π}, such that the input elements are confined to Z_{π} and the boundary function

$$\Delta : C_{\pi+1} \rightarrow C_{\pi}$$

has an image B_{π} which is contained in Z_{π}. This property of Δ is equivalent to saying that it reduces dimensions by 1 and satisfies

$$\Delta\Delta = 0.$$

This defines a chain complex. The following notation is taken for granted in the context of (C, Δ):

$$B = \text{image of } \Delta,$$

$$Z = \text{kernel of } \Delta,$$

$$H = Z/B,$$

$$D = C/Z.$$

DEFINITION I(3a). A *programming complex* is a chain complex (C, Δ) of the following form: C is a finitely generated free Abelian group, which reduces to a direct sum

$$C = C_1 + C_2 + \ldots + C_\mu,$$

and Δ is an endomorphism of C, whose restriction to each direct summand determines the sequence

$$0 \to C_\mu \to C_{\mu-1} \to \ldots \to C_2 \to C_1 \to 0$$

of homomorphisms, such that the composition $\Delta\Delta$ of two successive ones is zero. The complex is *free*, if H is free.

Any basis $\mathscr{C}$ for C in the above definition constitutes a *programming language* in (C, Δ), and the language is *free* if the complex is free. To characterize a language, it is necessary to know, not only the finite set $\mathscr{C}$, but also the function Δ, which gives an *interpretation* to each element of $\mathscr{C}$. That is, Δ lists for each x in $\mathscr{C}$ what it depends on and what it produces, if used in a program.

DEFINITION I(3b). In a free programming complex (C, Δ) a *program in C, which lists a basis for C*, consists of an *input*

$$h_1, h_2, \ldots, h_\eta,$$

taken from Z, and of *instructions*

$$d_1, d_2, \ldots, d_\beta$$

such that
 (i) for $\theta = 1, 2, \ldots, \beta$

$$\Delta(d_\theta) = -a_\theta + b_\theta,$$

where a_θ is in the group generated by

$$h_1, h_2, \ldots, h_\eta, d_1, d_2, \ldots, d_{\theta-1}, b_1, b_2, \ldots, b_{\theta-1}.$$

(ii) the elements

$$h_1, h_2, \ldots, h_\eta, d_1, d_2, \ldots, d_\beta, b_1, b_2, \ldots, b_\beta$$

constitute a basis for C.

(iii) η is the smallest value for which the above conditions are satisfied.

THEOREM I.5. *Let* (C, Δ) *be a free programming complex. Then,*

$$h_1, h_2, \ldots, h_\eta$$

and

$$d_1, d_2, \ldots, d_\beta$$

are the input and instructions, respectively, of a program in C, *which lists a basis for* C, *if, and only if,*

$$(h_1 + B), (h_2 + B), \ldots, (h_\eta + B)$$

and

$$(d_1 + Z), (d_2 + Z), \ldots, (d_\beta + Z)$$

are bases for H *and* D, *respectively.*

The proof of the above theorem is omitted since it is essentially the same as that of Theorem I.3.

I.7. COMPLEXES WHICH ARE NOT FREE

A programming complex is not free, if its homology group H has elements of finite order. In this case, if η is the rank of H, there is no program in C with η input elements, which will list a basis for C. However, there will be a program with η input elements which lists a mulitple of each element in a basis for C. This can be seen as follows: If the input only contains

$$h_1, h_2, \ldots, h_\eta,$$

whose cosets $(h_\theta + B)$ form a maximal linearly independent subset of H, and if H contains a non-zero coset $(h + B)$ such that for some integer τ greater than 1

$$\tau(h + B) = 0,$$

then h cannot be expressed as a linear combination of input elements and elements of B. Therefore, there is no program with the given input which lists h, and our only recourse is to place h on the list of input.

Accordingly, a program in a complex which is not free may be constructed by separating its homology group into a direct sum of a free group F and a group G of finite order. Then, an *input*

$$h_1, h_2, \ldots, h_\eta, b_1, b_2, \ldots, b_\phi$$

is determined by means of a basis

$$(h_1 + B), (h_2 + B), \ldots, (h_\eta + B)$$

for F and a minimal set

$$(b_1 + B), (b_2 + B), \ldots, (b_\phi + B)$$

of generators of G. For each b_θ in the input there is an integer $\tau_\theta > 1$ such that

$$\tau_\theta b_\theta \in B.$$

Hence, there exists d_θ such that

$$\Delta(d_\theta) = \tau_\theta b_\theta$$

for $\theta = 1, 2, \ldots, \phi$. Then a basis

$$(d_1 + Z), (d_2 + Z), \ldots, (d_\beta + Z)$$

is selected for D whose first ϕ terms are already determined by the input. This constructs a program.

   We conclude that the input required to list a basis for C is completely determined by the homology group H of the language. The elements of finite order in H cannot be ignored, unless we are willing to accept a program which lists a maximal linearly independent subset of C of the form

$$h_1, h_2, \ldots, h_\eta, d_1, d_2, \ldots, d_\beta, \tau_1 b_1, \tau_2 b_2, \ldots, \tau_\beta b_\beta,$$

which is not a basis unless all the τ's are equal to 1. In practice we use programs of this kind.

I.8. LANGUAGE ISOMORPHISMS

In order to put the problem of finding an algorithm in mathematical terms, it has been narrowed to a problem of programming. That is, the whole question of how to go about choosing a suitable programming language is avoided by assuming that a language $\mathscr{C}$ is given. The problem then becomes one of finding a program in $\mathscr{C}$, which lists the elements of $\mathscr{C}$. In the chain complex (C, Δ), uniquely determined by $\mathscr{C}$, it is easier to find a program, than in $\mathscr{C}$, because it is an infinite extension of $\mathscr{C}$ and offers greater programming freedom. The resulting program, however, lists a basis $\mathscr{C}'$ of C, which, in general, differs from $\mathscr{C}$. This can be corrected by finding a new complex (C, Γ) isomorphic to (C, Δ), such that a program in (C, Γ) which lists $\mathscr{C}$ will correspond under the isomorphism to the known program in (C, Δ) which lists $\mathscr{C}'$.

An isomorphism

$$\Psi : (C, \Gamma) \to (C, \Delta)$$

of one complex into another is known as a *chain isomorphism*. It can be characterized algebraically as follows: For any π, if Ψ is restricted to C_π, it is a group automorphism, and

$$\Psi\Gamma = \Delta\Psi.$$

Let us now verify that this equation is exactly what we would expect for two isomorphic languages $\mathscr{C}$ and $\mathscr{C}'$, when the isomorphism is expressed by the one-to-one correspondence

$$\Psi : \mathscr{C} \to \mathscr{C}'.$$

If x and $\Psi(x)$ are corresponding instructions in the two languages, then their interpretations, $\Gamma(x)$ and $\Delta(\Psi(x))$, respectively, must also correspond under Ψ. Therefore,

$$\Psi(\Gamma(x)) = \Delta(\Psi(x)).$$

Accordingly, it is appropriate to describe Ψ as a *language isomorphism*.

I.9. TO FIND AN ALGORITHM

Summarizing, let us see how the ideas of this chapter can be used to reduce the problem of finding a list-making algorithm to a general procedure.

STEP 1. Choose a language $\mathscr{C}$, such that it constitutes a basis for a chain complex (C, Δ). Choosing a language suitably is part of finding an algorithm, but from a mathematical viewpoint it is a procedure by which the statement of the problem is made complete. It provides a mathematical structure which will be examined in the remaining steps.

STEP 2. Find a program in (C, Δ) by means of Theorem I.5. Its input and instructions,

$$h_1, h_2, \ldots, h_\eta \text{ and } d_1, d_2, \ldots, d_\beta$$

can be any complete set of representatives of basis elements for the groups, (kernel Δ)/(image Δ) and C/(kernel Δ), associated with (C, Δ). The simplest program $\mathscr{P}'$ with this input and these instructions is the one whose output is

$$\Delta(d_1), \Delta(d_2), \ldots, \Delta(d_\beta).$$

This $\mathscr{P}'$ determines a basis $\mathscr{C}'$ for the group C, where $\mathscr{C}'$ differs, in general, from the desired basis $\mathscr{C}$. Also, $\mathscr{P}'$ is not in a form which is convenient for computation.

STEP 3. Find a complex (C, Γ), which is chain isomorphic to (C, Δ), such that the isomorphism matches program $\mathscr{P}'$ with a program $\mathscr{P}$, which is graphical and lists the elements of $\mathscr{C}$.

In practice this isomorphism is constructed from two isomorphisms, both defined on an intermediate complex (C, Ω), which has a particularly easy structure; i.e.

$$(C, \Delta) \cong (C, \Omega) \cong (C, \Gamma).$$

When this construction is carried out in subsequent chapters, it will be seen that each of the three complexes is simplest in its own way. (C, Δ) is algebraic and the easiest to define; (C, Ω) is combinatorial and has an easy structure to visualize; (C, Γ) is graphical and is the only practical one from a computational viewpoint.

STEP 4. Specify the program $\mathscr{P}$ inductively. This requires a generalization of the method described in Section I.3, where we were required only to list the elements of $\mathscr{C}_1$ by means of instructions taken from $\mathscr{C}_2$. First partition the entire language $\mathscr{C}$ into

$$\mathscr{C}_1 \cup \mathscr{C}_2 \cup \ldots \cup \mathscr{C}_\pi \cup \ldots \cup \mathscr{C}_\mu.$$

Then, for all π, taken in increasing order, a program for listing the elements of $\mathscr{C}_\pi$ can be specified in accordance with Section I.3, provided that the input includes, not only all input elements of dimension π, but also the list of all instructions which have already been determined in the listing of the elements of $\mathscr{C}_{\pi-1}$.

Flowchart 2 in the appendix specifies the program $\mathscr{P}$. It is an iteration of flowchart 1.

EXAMPLE I.6. Let the four above steps be illustrated by using them to determine an algorithm which lists every integer from 1 to μ, using only the kind of instruction which adds two integers together. It is obvious that this list of integers can be generated by starting with the integer 1 and adding 1 to it $\mu - 1$ times, but let us ignore this and see what algorithm we get by the method of this chapter. It will list, not only the integers 1 through μ and the instructions used to produce them, but also every other element of a suitable programming language.

STEP 1. Let us choose a language $\mathscr{C}$, based solely on the kind of instruction which tells us to take the sum of two integers, written consecutively. $\mathscr{C}$ must include, first of all, the set $\mathscr{C}_1$ of integers, being listed.

$$\mathscr{C}_1 = \{1, 2, \ldots, \mu\}$$

Next, it must include a set $\mathscr{C}_2$ of instructions which call for the addition of an ordered pair of integers in $\mathscr{C}_1$, whose sum is in $\mathscr{C}_1$.

$$\mathscr{C}_2 = \{(j_1, j_2) \in \mathscr{C}_1 \times \mathscr{C}_1 : (j_1 + j_2) \in \mathscr{C}_1\}.$$

Here, the instruction (j_1, j_2) depends on j_1 and j_2, and it produces $(j_1 + j_2)$. This is formalized by writing

$$\Delta((j_1, j_2)) = -j_1 - j_2 + (j_1 + j_2),$$

where ordinary integer addition is used inside the brackets and the formal addition of a free Abelian group outside.

Next we need a set $\mathscr{C}_3$ of instructions which depend on and produce elements of $\mathscr{C}_2$, in such a way that for any $x \in \mathscr{C}_3$

$$\Delta(\Delta(x)) = 0,$$

which is the chain complex property. This means that $\Delta(x)$ must be chosen in such a way as to be in the kernel of Δ. There are only two obvious choices:

$$(j_1, j_2) - (j_2, j_1)$$

and

$$-(j_2, j_3) + ((j_1 + j_2), j_3) - (j_1, (j_2 + j_3)) + (j_1, j_2).$$

We dismiss the first, because it equals a linear combination of elements of the second type. Accordingly, since three letters are involved, we define $\mathscr{C}_3$ to consist of elements of the form

$$p = (j_1, j_2, j_3),$$

and $\Delta(p)$ equals the 4-termed expression above. Now, it is obvious how to define an element p of $\mathscr{C}_\pi$ for $\pi > 3$. It is π-termed sequence of positive integers

$$p = (j_1, j_2, \ldots, j_\pi),$$

and

$$\Delta(p) = -(j_2, \ldots, j_\pi)$$
$$+ \sum_{\theta=1}^{\pi-1} (-1)^\theta (j_1, \ldots, (j_\theta + j_{\theta+1}), \ldots, j_\pi)$$
$$+ (-1)^\pi (j_1, \ldots, j_{\pi-1}).$$

These instructions involve many terms, but each of them only requires the addition of two integers. This is another reason for avoiding the first choice, suggested above for $\Delta(p)$, that is,

$$(j_1, j_2) - (j_2, j_1).$$

If $\Delta(p)$ took such a value, then p would be a transposition instruction and an unnecessary enlargement of the language.

This completes the choice of language, which is the most delicate part of finding an algorithm. The remaining steps are straightforward in the sense of being answers to well-stated mathematical questions.

STEP 2. A program $(\mathscr{H}, \mathscr{D})$ in (C, Δ) is as follows:
$\mathscr{H}$ equals the singleton set $\{(1)\}$, and $\mathscr{D}$ equals the set $\{(1, j_2, j_3, \ldots, j_\pi): \pi \geq 2$ and $(1 + j_2 + j_3 + \ldots + j_\pi) \leq \mu\}$ of all elements of $\mathscr{C}$, whose first term is 1, and having two or more terms.

To prove that this constitutes a program of the required kind, it is only necessary to show that the set

$$\mathscr{H} \cup \mathscr{D} \cup \Delta \mathscr{D}$$

is a basis for C. This follows easily from the observation that

$$\mathscr{C} = \mathscr{H} \cup \mathscr{D} \cup \mathscr{B},$$

in which $\mathscr{B}$ is formed by replacing each element

$$\Delta((1, j_2, j_3, \ldots, j_\pi))$$

of $\Delta \mathscr{D}$ by the single term

$$((1 + j_2), j_3, \ldots, j_\pi).$$

STEP 3. A new program $\mathscr{P}$ is found as follows: Replace Δ by a boundary operation Γ defined as follows on elements of $\mathscr{D}$:

$$\Gamma((1, j_2, j_3, \ldots)) = -(j_2, j_3, \ldots) + ((1 + j_2), j_3, \ldots).$$

This, combined with $\Gamma((1)) = 0$ and $\Gamma\Gamma = 0$, determines a chain complex (C, Γ), which is isomorphic to (C, Δ). The new program $\mathscr{P}$ in (C, Γ) corresponds under the isomorphism to the old program $(\mathscr{H}, \mathscr{D})$ in (C, Δ).

In order to prove the isomorphism, write it as

$$\Psi : (C, \Gamma) \to (C, \Delta),$$

where Ψ is the identity on $\mathscr{D} \cup \mathscr{H}$, and its values on $\Gamma \mathscr{D}$ are determined by the condition $\Psi\Gamma = \Delta\Psi$. The proof that Ψ is a chain complex isomorphism follows easily from the fact that

$$\mathscr{D} \cup \mathscr{H} \cup \Gamma \mathscr{D}$$

is a basis for C and that the condition $\Psi\Gamma = \Delta\Psi$ holds on all elements of C.

Instructions				Input and Output
				The input is (1)
$\mathscr{L}(1)$	$= (1, 1),$	which produces		(2)

(The above line means the same as $\Gamma((1, 1)) = -(1) + (2)$.)

$\mathscr{L}(2)$	$= (1, 2)$	which produces		(3)
$\mathscr{L}(3)$	$= (1, 3)$	which produces		(4)
$\mathscr{L}(4)$	$= $ none			
$\mathscr{L}(1, 1)$	$= (1, 1, 1)$	which produces		(2, 1)
$\mathscr{L}(1, 2)$	$= (1, 1, 2)$	which produces		(2, 2)
$\mathscr{L}(1, 3)$	$= $ none			
$\mathscr{L}(2, 1)$	$= (1, 2, 1)$	which produces		(3, 1)
$\mathscr{L}(2, 2)$	$= $ none			
$\mathscr{L}(3, 1)$	$= $ none			
$\mathscr{L}(1, 1, 1)$	$= (1, 1, 1, 1)$	which produces		(2, 1, 1)
$\mathscr{L}(1, 1, 2)$	$= $ none			
$\mathscr{L}(1, 2, 1)$	$= $ none			
$\mathscr{L}(2, 1, 1)$	$= $ none			
$\mathscr{L}(1, 1, 1, 1)$	$= $ none			

STEP 4. The program $\mathscr{P}$, which was determined mathematically in the last step, now has to be specified in a way which is suited to computation. One way to do this is to introduce a subroutine $\mathscr{L}$, as described in Section I.3. $\mathscr{L}$ is here defined as follows: For any element p in $\mathscr{C}$,

$$p = (j_1, j_2, \ldots),$$

such that $j_1 + j_2 + j_3 + \ldots < \mu,$

$$\mathscr{L}(p) = (1, j_1, j_2, \ldots),$$

and, if the sum of the terms of p are equal to μ, then $\mathscr{L}(p)$ is not defined. If $\mathscr{L}(p)$ exists, then it is the instruction which depends on $(j_1, j_2, \ldots)$ and produces $(1 + j_1, j_2, \ldots)$.

To demonstrate the use of $\mathscr{L}$, let us consider the case when $\mu = 4$. The elements of $\mathscr{C}$ are displayed in two columns in the above table. $\mathscr{L}$ is applied to each new element after it is listed. Thus, a complete list of $\mathscr{C}$ is produced, where $\mathscr{C}$ is what is known as the set of all *compositions* of the integer 4.

Chapter II

Monomials

II.1. THE LANGUAGE

This chapter is concerned with how to make a list of monomials by means of multiplication instructions. The *language of monomials* is of the form

$$\mathscr{C} = \mathscr{C}_1 \cup \mathscr{C}_2 \cup \ldots \cup \mathscr{C}_\pi \cup \ldots \cup \mathscr{C}_\rho,$$

where

(i) $\mathscr{C}_1$ is the set of all monomials of degree at least 1, which divide a given monomial m of the form

$$m = a_1^{\rho_1} a_2^{\rho_2} \ldots a_\mu^{\rho_\mu},$$

the *degree* of m being ρ, where

$$\rho = \rho_1 + \rho_2 + \ldots + \rho_\mu.$$

(ii) $\mathscr{C}_\pi$ is the set consisting of every π-termed sequence of monomials of the form

$$p = j_1 \times j_2 \times \ldots \times j_\pi,$$

such that the product $j_1 j_2 \ldots j_\pi$ of its terms is an element of $\mathscr{C}_1$. As a multiplication instruction, p will be given three different, but isomorphic, interpretations Δ, Ω, and Γ in the next three sections.

The interpretation Δ of the elements of $\mathscr{C}$, as instructions, is defined by embedding $\mathscr{C}$ in a chain complex (C, Δ). It is the easiest interpretation to define, being based on an algebraic formula. Ω is a more complicated combinatorial interpretation, but conceptually the simplest, since it reduces to a map of $\mathscr{C}$ into itself, so that the extension of the language to a group C becomes unessential. The properties common to all three interpretations of $\mathscr{C}$ are derived in the context of (C, Ω), and then they are translated into properties of (C, Δ) and (C, Γ) by chain isomorphisms. Γ is a graph theoretic interpretation, by means of which a program for listing the elements of $\mathscr{C}$ can be specified inductively, that is, reduced to flowchart 2 of the appendix.

II.2. THE INTERPRETATION Δ

An interpretation of $\mathscr{C}$ is defined by embedding it in a chain complex (C, Δ). C is the free Abelian group generated by $\mathscr{C}$, and Δ is determined by the following formula:

$$\Delta(j_1 \times j_2 \times \ldots \times j_\pi) = -(j_2 \times \ldots \times j_\pi) -$$
$$- \sum_{\theta=1}^{\pi-1} (-1)^\theta (j_1 \times \ldots \times j_\theta j_{\theta+1} \times \ldots \times j_\pi) -$$
$$- (-1)^\pi (j_1 \times \ldots \times j_{\pi-1}),$$

where $j_1 \times \ldots \times j_\pi$ is any element of $\mathscr{C}_\pi$.

The above formula is an obvious extension of

$$\Delta(j_1 \times j_2) = -j_1 - j_2 + j_1 j_2,$$

in which $j_1 \times j_2$ is an instruction which depends on the pair j_1, j_2 of monomials and produces their product $j_1 j_2$. This kind of instruction defines an elementary language $(\mathscr{C}_1, \mathscr{C}_2)$ in the sense of Definition I.1 $(\nu = 2)$.

As an example of an instruction in the full language $\mathscr{C}$, let us consider what interpretation Δ places on an element $j_1 \times j_2 \times j_3$ of $\mathscr{C}_3$. Applying Δ to it, we find that it depends on a two-step multiplication,

$$a = j_2 \times j_3 + j_1 \times j_2 j_3$$

and produces another two-step multiplication

$$b = j_1 \times j_2 + j_1 j_2 \times j_3.$$

a and b both depend on $j_1 + j_2 + j_3$ and both produce $j_1 j_2 j_3$. As instructions they are different, since they perform the multiplications differently, but they are homologous, having the same interpretation; that is,

$$\Delta(a) = \Delta(b),$$

and this is equivalent to the associative law for multiplication of monomials. This equation can be written as follows:

$$0 = \Delta(-a + b) = \Delta\Delta(j_1 \times j_2 \times j_3).$$

This proves that $\Delta\Delta = 0$ for dimension 3, and it is easily verified in general, which justifies our definition of (C, Δ) as a chain complex.

II.3. THE INTERPRETATION Ω

A chain complex (C, Ω), isomorphic to (C, Δ), and having $(\mathscr{H}, \mathscr{D}, \mathscr{B})$ as a standard basis, is constructed as follows:

Let the *prime factors* $a_1, a_2, \ldots, a_n$ of m, that is, the monomials of first degree, have an arbitrary linear ordering, which remains fixed throughout this discussion. For any p in $\mathscr{C}_\pi$, where

$$p = j_1 \times j_2 \times \ldots \times j_\pi,$$

let θ be the largest subscript such that

$$j_1, j_2, \ldots, j_\theta$$

is a properly decreasing sequence of first degree monomials. Then

(i) $p \in \mathscr{H}$, if $\theta = \pi$. Hence, if p is not in $\mathscr{H}$ and θ exists, then $\theta < \pi$.

(ii) $p \in \mathscr{D}$, if θ exists and j_θ is less than or equal to the smallest prime factor in $j_{\theta+1}$.

(iii) $p \in \mathscr{B}$, if θ does not exist, or j_θ is greater than the smallest prime factor in $j_{\theta+1}$.

The above construction allows us to characterize the subscript θ in a new and useful way: Let any p in $\mathscr{D}_\pi$ be expressed in the form

$$j_1 \times \ldots \times j_\pi;$$

then $j_1 \times \ldots \times j_\theta$ is the *homology part of p*, if θ is the largest subscript such that

$$j_1 \times j_2 \times \ldots \times j_\theta \in \mathscr{H}.$$

The *boundary function* Ω is defined as zero on all elements of $\mathscr{B} \cup \mathscr{H}$, and

$$\Omega(j_1 \times \ldots \times j_\pi) = j_1 \times \ldots \times j_\theta j_{\theta+1} \times \ldots \times j_\pi$$

for each $j_1 \times \ldots \times j_\pi$ in $\mathscr{D}$, whose homology part is $j_1 \times \ldots \times j_\theta$.

The expression on the right-hand side of the equation is in $\mathscr{B}$, because $j_{\theta-1}$ is greater than the smallest prime factor j_θ of the monomial $j_\theta j_{\theta+1}$. Therefore,

$$\Omega\Omega \; = \; 0.$$

It follows from this that the map Ω from $\mathscr{C}$ into $\mathscr{C} \cup \{0\}$, when it is extended to a linear transformation of C into itself, determines a unique chain complex, having $(\mathscr{H}, \mathscr{D}, \mathscr{B})$ as a standard basis.

II.4. THE INTERPRETATION Γ

A chain complex (C, Γ) is now constructed, which is isomorphic to both (C, Δ) and (C, Ω). The main fact about this complex is that it contains a *graphical program*, which lists without repetitions the elements of $\mathscr{C}$. $\mathscr{H}$ and $\mathscr{D}$, as defined in the last section, are the input and instructions of such a program.

$\mathscr{H}$ and $\mathscr{D}$ are defined in Section II.3. If p is a π-dimensional element of $\mathscr{D}$ of the form

$$p \; = \; j_1 \times \ldots \times j_\pi,$$

having $j_1 \times \ldots \times j_\theta$ as its homology part, then

$$\Gamma(p) \; = \; -(j_1 \times \ldots \times \hat{j}_\theta \times \ldots \times j_\pi) +$$
$$+ \, (j_1 \times \ldots \times j_\theta j_{\theta+1} \times \ldots \times j_\pi)$$

where $\hat{j}_\theta$ is understood to mean that $j_\theta \times$ has been deleted. The set

$$\mathscr{H} \cup \mathscr{D} \cup \Gamma \mathscr{D}$$

constitutes a basis for C. Therefore, a chain complex (C, Γ) is completely determined by requiring Γ to take the value zero on all elements of $\mathcal{H} \cup \Gamma \mathcal{D}$.

II.5. LANGUAGE ISOMORPHISMS

There is an obvious chain isomorphism from (C, Ω) to (C, Γ). It is the identity map, when it is restricted to $\mathcal{H} \cup \mathcal{D}$, and it takes each element $\Omega(d)$ of $\mathcal{B}$ to $\Gamma(d)$.

The remainder of this section is devoted to constructing a chain isomorphism

$$\Psi : (C, \Omega) \to (C, \Delta).$$

It is a group automorphism of C, such that

$$\Psi\Omega \ = \ \Delta\Psi.$$

Ψ maps the standard basis $(\mathcal{H}, \mathcal{D}, \mathcal{B})$ of (C, Ω) to a standard basis $(\mathcal{H}', \mathcal{D}', \mathcal{B}')$ of (C, Δ), and it transforms the program in (C, Ω) with input $\mathcal{H}$ and instructions $\mathcal{D}$, which lists the elements of $\mathcal{C}$, into a program in (C, Δ) with input $\mathcal{H}'$ and instructions $\mathcal{D}'$, which lists the element of $\mathcal{C}'$, where

$$\mathcal{C}' \ = \ \mathcal{H}' \cup \mathcal{D}' \cup \mathcal{B}'.$$

DEFINITION II.1. For

$$p = j_1 \times \ldots \times j_\theta \in \mathcal{B}_\theta$$

and

$$q = j_{\theta+1} \times \ldots \times j_\pi \in \mathcal{B}_{\pi-\theta}$$

the *shuffle product* of p and q is an element of $\mathcal{C}_\pi$ defined as follows:

$$p \wedge q = \sum \kappa(k_1 \times \ldots \times k_\pi),$$

where the sum is taken over every $k_1 \times \ldots \times k_\pi$, formed by permuting the terms of $j_1 \times \ldots \times j_\pi$, in such a way as to preserve separately the linear orderings of the terms of $j_1 \times \ldots \times j_\theta$ and those of $j_{\theta+1} \times \ldots \times j_\pi$, and κ is $+1$ or -1 according as the permutation is even or odd.

Notice in this definition that each $k_1 \times \ldots \times k_\pi$ is a shuffle, as with playing cards, of $j_1 \times \ldots \times j_\theta$ with $j_{\theta+1} \times \ldots \times j_\pi$, and that the sum is taken over all such shuffles. The shuffle is distributive in the sense that if $p_1 \wedge q$ and $p_2 \wedge q$ are defined, then we can write their sum as

$$(p_1 + p_2) \wedge q.$$

LEMMA II.1. *A chain isomorphism*

$$\Psi : (C, \Omega) \rightarrow (C, \Delta)$$

which maps a standard basis ($\mathscr{H}$, $\mathscr{D}$, $\mathscr{B}$) *into another one* ($\mathscr{H}'$, $\mathscr{D}'$, $\mathscr{B}'$) *is constructed as follows:*
   (i) *For each element of* $\mathscr{H}$ *of dimension* π

$$\Psi(j_1 \times \ldots \times j_\pi) = j_1 \wedge \ldots \wedge j_\pi.$$

   (ii) *For each element of* $\mathscr{D}$ *of dimension* π

$$\Psi(j_1 \times \ldots \times j_\pi) = j_1 \wedge \ldots \wedge j_{\theta-1} \wedge (j_\theta \times \ldots \times j_\pi)$$

where $j_1 \times \ldots \times j_\theta$ is the homology part.
   (iii) *For each element of* $\mathscr{B}$, *expressed as* $\Omega(d)$,

$$\Psi(\Omega(d)) = \Delta(\Psi(d)).$$

Proof. This lemma holds, if a linear ordering of the elements of $\mathscr{C}$ can be found with the following property: For every p in $\mathscr{C}$, if $\Psi(p)$ is expressed as a linear combination of elements of $\mathscr{C}$ with non-zero coefficients, then p is the largest (with respect to the linear order) of these elements, and its coefficient is $+1$ or -1. This condition is equivalent to saying that Ψ is a change of basis, which is expressible by a square matrix with units (± 1) on the diagonal and zeros above it.

Let $\mathscr{C}$ be linearly ordered such that $(j_1 \times j_2 \times \ldots)$ is less than $(k_1 \times k_2 \times \ldots)$, whenever

(i) the degree of monomial $(j_1 j_2 \ldots)$ is less than that of mon-omial $(k_1 k_2 \ldots)$, and,

(ii) if the degrees of $(j_1 j_2 \ldots)$ and $(k_1 k_2 \ldots)$ are equal, then the order is lexicographic, as described in the next paragraph:

Here, a double lexicographic ordering is needed. First, the mon-omials are ordered by writing them as non-decreasing sequences of prime factors and ordering these sequences lexicographically, rela-tive to the ordered set of primes. Next,

$$j_1 \times j_2 \times \ldots \quad \text{and} \quad k_1 \times k_2 \times \ldots$$

are ordered by regarding them as sequences of monomials and ordering these sequences lexicographically, relative to the ordered set of monomials.

For elements of $\mathscr{D} \cup \mathscr{H}$ it is obvious that p appears in $\Psi(p)$ with a coefficient of $+1$ or -1, and that all other terms (elements of $\mathscr{C}$ with non-zero coefficients) are smaller than p in the linear ordering. Suppose $p \in \mathscr{B}$; then it is of the form

$$p = \Omega(j_1 \times \ldots \times j_{\pi+1}) \in \mathscr{C}_\pi$$

where $j_1 \times \ldots \times j_\theta$ is the homology part and j_θ is less than or equal to the smallest prime factor in $j_{\theta+1}$.

$$\Psi(p) = \Psi\Omega(j_1 \times \ldots \times j_{\pi+1}) = \Delta\Psi(j_1 \times \ldots \times j_{\pi+1})$$
$$= \Delta(j_1 \wedge \ldots \wedge j_{\theta-1} \wedge (j_\theta \times j_{\theta+1} \times \ldots \times j_{\pi+1}))$$
$$= (-1)^{\theta-1}(j_1 \wedge \ldots \wedge j_{\theta-1} \wedge \Delta(j_\theta \times \ldots \times j_{\pi+1})).$$

The last step can be verified directly or as a corollary of the general rule

$$\Delta(s \wedge t) = \Delta(s) \wedge t + (-1)^\sigma s \wedge \Delta(t),$$

where $s \in \mathscr{C}_\sigma$, $t \in \mathscr{C}_\tau$, and $s \times t \in \mathscr{C}_{\sigma+\tau}$. In its last form $\Psi(p)$ is seen to have the desired property, so that Ψ is a group automorphism.

Finally, Ψ is a chain automorphism, because it has been constructed in such a way that

$$\Psi\Omega = \Delta\Psi.$$

This proves the lemma, and also it proves the following theorem.

THEOREM II.1. *A standard basis* $(\mathscr{H}', \mathscr{D}', \mathscr{B}')$ *for the monomial complex* (C, Δ) *is as follows: Let the first degree monomials have a linear ordering assigned to them; then* $\mathscr{H}'$ *is the set of all*

$$j_1 \wedge j_2 \wedge \ldots \wedge j_\pi,$$

where $j_1, j_2, \ldots, j_\pi$ *is a properly decreasing sequence of first degree monomials.* $\mathscr{D}'$ *is the set of all*

$$j_1 \wedge j_2 \wedge \ldots j_{\theta-1} \wedge (j_\theta \times j_{\theta+1} \times \ldots \times j_\pi),$$

such that $j_1, j_2, \ldots, j_\theta$ *is a properly decreasing sequence of first degree monomials, and* $j_{\theta+1}$ *is a monomial whose first degree factors are all greater than or equal to* j_θ $(1 \leqslant \theta < \pi)$. $\mathscr{B}'$ *is the set of all* $\Delta(d)$, *where* d *is an element of* $\mathscr{D}'$.

II.6. A PROGRAM FOR MONOMIALS

The groups H and D are now completely determined for each of the complexes (C, Δ), (C, Ω), and (C, Γ). Therefore, we have determined the set of all programs in each complex, which list a basis for C.

In particular, $\mathscr{H}$ and $\mathscr{D}$ are the input and instructions of a graphical program in (C, Γ) which lists without repetitions the elements of $\mathscr{C}$. The instructions have to be ordered in such a way that the element which each instruction depends on is in $\mathscr{H}$ or was produced by a previous instruction, which can be accomplished by any ordering such that

$$j_1 \times j_2 \times \ldots \text{ precedes } k_1 \times k_2 \times \ldots,$$

whenever the degree of the monomial $j_1 j_2 \ldots$ is less than that of the monomial $k_1 k_2 \ldots$. Therefore, one way to specify the program is to give $\mathscr{H}$ explicitly in the form of a list without repetitions, and to give $\mathscr{D}$ explicitly in the form of an increasing sequence relative to the above type of linear ordering.

A more useful way to specify the program is by induction, as described in Section I.3. This means finding a subroutine $\mathscr{L}$, which assigns to each element x of $\mathscr{C}$ the list $\mathscr{L}(x)$ of all the instructions which depend on x. Then the program is completely specified by substituting this $\mathscr{L}$ into flowchart 2 of the appendix. It is immediately seen from the definition of Γ in Section II.4 how the subroutine $\mathscr{L}$ is to be constructed. For any x in $\mathscr{C}$ it is easy to list the set $\mathscr{L}(x)$ of all p in $\mathscr{C}$, such that x will be the negative term in $\Gamma(p)$. A subroutine to do this is given by part L of flowchart 4 in the appendix. An entire program in (C, Γ), for listing the elements of $\mathscr{C}$, is specified by flowcharts 2 and 4 of the appendix, taken together. The part of this program which merely lists the set $\mathscr{C}_1$ of monomials (not the whole language) is specified by flowcharts 1 and 3 of the appendix, taken together.

The programs of this chapter have $2^\mu - 1$ input elements, which is a minimum relative to the particular kinds of instructions allowed. It is still a minimum when we are allowed to use any two adjacent terms in any element of $\mathscr{C}$. This is proved in the next theorem, and it shows that we have not hampered our languages by the special ways in which we have constructed them.

THEOREM II.2. *Let $\mathscr{C}$ be a monomial language in μ indeterminates (first degree monomials), and let $\mathscr{H}$, $\mathscr{D}$, $\mathscr{B}$ be any partition of the set $\mathscr{C}$ into disjoint subsets, such that there exists a one-to-one correspondence Θ of $\mathscr{D}$ onto $\mathscr{B}$ of the form*

$$\Theta(j_1 \times j_2 \times \ldots \times j_\pi) = (j_1 \times j_2 \times \ldots \times j_\phi j_{\phi+1} \times \ldots \times j_\pi),$$

where ϕ varies with the elements of $\mathscr{D}$ and

$$1 \leqslant \phi < \pi.$$

Under these conditions, if Θ is chosen so that $\mathscr{D}$ and $\mathscr{B}$ are large as possible, then $\mathscr{H}$ has $2^\mu - 1$ elements.

Proof. Let $\mathscr{C}$ be partitioned into a family of subsets, where each subset $\mathscr{C}(j)$ corresponds to a single element j of $\mathscr{C}_1$:

$$\mathscr{C}(j) = \{(j_1 \times j_2 \times \ldots) : (j_1 j_2 \ldots) = j\}.$$

Since p and $\Theta(p)$ are always in the same subset, the number η of elements, which are not paired up by Θ, is given by

$$\eta = \sum_j \eta(j),$$

where $\eta(j)$ is the number of unpaired elements in $\mathscr{C}(j)$.

In any finite set the absolute minimum of unpairable elements is 0 or 1, according as the number of elements in the set is even or odd. This minimum is attained by the pairing Ω, applied to $\mathscr{C}(j)$, because the set $\mathscr{H}$ of unpaired elements in $\mathscr{C}$ never has more than one of its elements in any subset $\mathscr{C}(j)$. Since $\mathscr{H}$ has $2^\mu - 1$ elements relative to the pairing Ω, the same is true relative to Θ. This proves the theorem.

II.7. SAMPLE PROGRAMS

EXAMPLE II.1. Use a program in (C, Γ) to list all the factors of the monomial m, when

$$m = ab^2c.$$

The program depends on the prime factors of m have a linear ordering, for which we use the alphabetic ordering. The given input is the list

$$a, b, c.$$

We use flowchart 1 of the appendix. Each line below represents one passage through the loop, and the arrow $\rightarrow$ runs from each instruction to the element it produces. The right-hand column is the desired list.

$$\text{Input} = \begin{cases} a \\ b \\ c \end{cases}$$

$$\mathscr{L}(a) = \phi$$

$$\mathscr{L}(b) = \begin{cases} a \times b \to ab \\ b \times b \to b^2 \end{cases}$$

$$\mathscr{L}(c) = \begin{cases} a \times c \to ac \\ b \times c \to bc \end{cases}$$

$$\mathscr{L}(ab) = \phi$$

$$\mathscr{L}(b^2) = a \times b^2 \to ab^2$$

$$\mathscr{L}(ac) = \phi$$

$$\mathscr{L}(bc) = \begin{cases} a \times bc \to abc \\ b \times bc \to b^2c \end{cases}$$

$$\mathscr{L}(ab^2) = \phi$$

$$\mathscr{L}(abc) = \phi$$

$$\mathscr{L}(b^2c) = a \times b^2c \to ab^2c.$$

EXAMPLE II.2. Use a program in (C, Γ) to list the entire monomial language for the case when

$$m = a^3b.$$

The input $\mathscr{H}$ is the list

$$a, b, b \times a.$$

We follow flowchart 2 of the appendix. Each line below represents one passage through the inner loop of the flow chart, but no line is written, when $\lambda = 0$. The lines are indented, in such a way that the elements of $\mathscr{C}_1, \mathscr{C}_2, \mathscr{C}_3,$ and $\mathscr{C}_4$ will appear in separate columns, one for each passage through the outer loop. An arrow $\to$ runs from each instruction to the element it produces.

$\mathcal{E}_1$	$\mathcal{E}_2$	$\mathcal{E}_3$	$\mathcal{E}_4$

$$\mathcal{H}_1 = \begin{cases} a \\ b \end{cases}$$

$\quad\to a^2$

$\quad\to ab$

$\quad\to a^3$

$\quad\to a^2b$

$\quad\to a^3b$

$\mathcal{E}_2$:

$$\mathcal{L}(a) = a \times a$$
$$\mathcal{L}(b) = a \times b$$
$$\mathcal{L}(a^2) = a \times a^2$$
$$\mathcal{L}(ab) = a \times ab$$
$$\mathcal{L}(a^2b) = a \times a^2b$$
$$\mathcal{H}_2 = b \times a$$

$\quad\to a^2 \times a$

$\quad\to a^2 \times b$

$\quad\to a^2 \times ab$

$\quad\to ab \times a$

$\quad\to b \times a^2$

$\quad\to a^3 \times b$

$\quad\to a^2b \times a$

$\quad\to ab \times a^2$

$\quad\to b \times a^3$

$\mathcal{E}_3$:

$$\mathcal{L}(a \times a) = a \times a \times a$$
$$\mathcal{L}(a \times b) = a \times a \times b$$
$$\mathcal{L}(a \times ab) = a \times a \times ab$$
$$\mathcal{L}(b \times a) = \begin{cases} a \times b \times a \\ b \times a \times a \end{cases}$$
$$\mathcal{L}(a^2 \times b) = a \times a^2 \times b$$
$$\mathcal{L}(ab \times a) = a \times ab \times a$$
$$\mathcal{L}(b \times a^2) = \begin{cases} a \times b \times a^2 \\ b \times a \times a^2 \end{cases}$$

$\quad\to a^2 \times a \times b$

$\quad\to a^2 \times b \times a$

$\quad\to ab \times a \times a$

$\quad\to b \times a^2 \times a$

$\mathcal{E}_4$:

$$\mathcal{L}(a \times a \times b) = a \times a \times a \times b$$
$$\mathcal{L}(a \times b \times a) = a \times a \times b \times a$$
$$\mathcal{L}(b \times a \times a) = \begin{cases} a \times b \times a \times a \\ b \times a \times a \times a \end{cases}$$

Chapter III

Factorizations

III.1. THE LANGUAGE

This chapter is concerned with how to make a list of the factorizations of a monomial m, by means of instructions which change
one factorization into another by multiplying two of its factors
together. We make use of the concepts of Chapter II on monomials,
because an element of the language of factorizations of m is equal
to a list of elements from the language of monomials. The monomial m, to be factored, is the same as m in Chapter II,

$$m = a_1^{\rho_1} a_2^{\rho_2} \ldots a_\mu^{\rho_\mu}.$$

The language $\mathscr{C}$ of factorizations of m and its interpretations Δ,
Ω, and Γ are analogous to the structures in Chapter II, which are
denoted by the same symbols. In other words, $\mathscr{C}$ will be embedded
in three chain complexes (C, Δ), (C, Ω), and (C, Γ), which are
adaptations of the complexes of the last chapter to a more elabor-
ate situation. In their essential character the three complexes are
algebraic, combinatorial, and graphical.

The *language of factorizations of the monomial m* is of the form

$$\mathscr{C} = \mathscr{C}_1 \cup \mathscr{C}_2 \cup \ldots \cup \mathscr{C}_\pi \cup \ldots \cup \mathscr{C}_\rho,$$

where

(i) $\mathscr{C}_1$ is the set consisting of every list

$$j_1, j_2, \ldots$$

of monomials, such that

$$j_1 j_2 \ldots = m.$$

The list is called a *factorization of m*, and its entries are called *factors*.

(ii) $\mathscr{C}_\pi$ is the set consisting of every list

$$(j_{11} \times j_{12} \times \ldots), (j_{21} \times j_{22} \times \ldots), \ldots$$

of elements of the language of monomials, in which the symbol $\times$ appears $\pi - 1$ times and

$$(j_{11} j_{12} \ldots)(j_{21} j_{22} \ldots) \ldots = m,$$

and such that no two even-dimensional entries on the list are identical. (Dimension equals the number of $\times$'s plus 1.)

The last condition of this construction, eliminating repeated even-dimensional factors in an element of $\mathscr{C}$, is made in anticipation of the fact that such terms will be meaningless as instructions under the interpretation Δ, which is given in the next section.

III.2. THE INTERPRETATION Δ

An interpretation of $\mathscr{C}$ is defined by embedding it in a chain complex (C, Δ). C is the free Abelian group generated by $\mathscr{C}$, on the assumption that each element of $\mathscr{C}$ has its list of factors written in a unique conventional order. In other words, each element of $\mathscr{C}$ is represented by a unique sequence

$$(\ldots, s, t, \ldots)$$

of elements of the language of monomials. Any other sequence, representing the same element of $\mathscr{C}$, belongs to the group C, but may differ by a sign, according to the following rule: If s and t are elements of dimensions σ and τ, respectively, in the language of monomials, then

$$(\ldots, t, s, \ldots) = (-1)^{(\sigma-1)(\tau-1)}(\ldots, s, t, \ldots).$$

This rule relies on the fact, already stated, that s and t cannot be equal if they are even-dimensional.

If a basis element of C is of dimension π, then the symbol $\times$ appears in it $(\pi-1)$ times, and it may be written in the form

$$p_1 \times p_2 \times \ldots \times p_\pi,$$

where $(p_1, p_2, \ldots, p_\pi)$ would be in $\mathscr{C}_1$. For $\pi > 1$ the boundary function Δ is defined by

$$\Delta(p_1 \times p_2 \times \ldots \times p_\pi)$$

$$= \sum_{\theta=1}^{\pi-1} (-1)^\theta [(p_1 \times \ldots \times p_\theta, p_{\theta+1} \times \ldots \times p_\pi) -$$

$$- (p_1 \times \ldots \times p_\theta p_{\theta+1} \times \ldots \times p_\pi)]$$

and, if $\pi = 1$, the value of Δ is zero.

It is easily verified that (C, Δ) is a complex by showing that

$$\Delta(\ldots, s, t, \ldots) = (-1)^{(\sigma-1)(\tau-1)}\Delta(\ldots, t, s, \ldots)$$

and that

$$\Delta\Delta = 0.$$

The above formula is an obvious extension of

$$\Delta(p_1 \times p_2) = -(p_1, p_2) + p_1 p_2,$$

where $p_1 \times p_2 \in \mathscr{C}_2$. Let p_1 and p_2 be expressed as sequences of monomials:

$$p_1 = j_1, j_2, \ldots, j_\kappa$$
$$p_2 = j_{\kappa+1}, j_{\kappa+2}, \ldots, j_\lambda.$$

Then $p_1 \times p_2$ depends on

$$j_1, j_2, \ldots, j_\lambda$$

and produces the element obtained from this by multiplying j_κ and $j_{\kappa+1}$ together. The instruction, written out in full, is

$$j_1, \ldots, j_{\kappa-1}, j_\kappa \times j_{\kappa+1}, j_{\kappa+2}, \ldots, j_\lambda;$$

it is independent of the order in which its $\lambda - 1$ factors are written. Only elements of higher dimension can change sign when their factors are permuted.

As an example of a higher dimensional instruction, consider

$$j_1 \times j_2, j_3 \times j_4.$$

Applying Δ to it we find that it depends on a pair of multiplications,

$$a = (j_1, j_2, j_3 \times j_4) + (j_1 \times j_2, j_3 j_4)$$

and produces another pair

$$b = (j_1 \times j_2, j_3, j_4) + (j_1 j_2, j_3 \times j_4).$$

a is homologous to b, since both depend on

$$j_1, j_2, j_3, j_4$$

and both produce

$$j_1 j_2, j_3 j_4.$$

Notice that the instruction can have its two parts transposed, but there is a sign change. That is,

$$(j_1 \times j_2, j_3 \times j_4) = -(j_3 \times j_4, j_1 \times j_2).$$

III.3. THE INTERPRETATION Ω

The results of Chapter II will now be used to construct a chain complex (C, Ω), which is essentially isomorphic to the complex (C, Δ) of the last section and has a standard basis denoted by $(\mathcal{H}, \mathcal{D}, \mathcal{B})$. The monomial language of Chapter II will be denoted by $\mathcal{C}(m)$, the monomial complex of Section II.3 by $(C(m), \Omega_m)$, and its standard basis by $(\mathcal{H}(m), \mathcal{D}(m), \mathcal{B}(m))$.

Elements of the factorization language $\mathcal{C}$ are finite lists of elements of $\mathcal{C}(m)$, which can be written uniquely, if a conventional linear ordering is adopted for them. Similarly, the elements of the monomial language $\mathcal{C}(m)$ are sequences of monomials, which can be written uniquely, if the monomials have their prime factors in non-decreasing order relative to the fixed linear ordering of the primes.

DEFINITION III.1. An element of the factorization language $\mathcal{C}$ is in *conventional form*, when it is expressed by a sequence of elements of the monomial language $\mathcal{C}(m)$, placed in non-decreasing order relative to a linear ordering of $\mathcal{C}(m)$, chosen as follows:

(i) The subset $\mathcal{H}(m) \cup \mathcal{D}(m)$ is arbitrarily ordered.

(ii) Each element $\Omega_m(d)$ of $\mathcal{B}(m)$ is the immediate predecessor or immediate successor of d, according as $\Omega_m(d)$ is of even or odd dimension.

In (i) of the above definition, let us assume that the arbitrary ordering is lexicographic, if nothing is said to the contrary. This ordering was defined on $\mathscr{C}(m)$ in the proof of Lemma II.1.

Let p be any element of $\mathscr{B}$ in conventional form. Then we can write

$$p = (s_1, s_2, \ldots, s_\theta, \ldots),$$

which is a non-decreasing sequence of elements of $\mathscr{C}(m)$.

(i) $p \in \mathscr{H}$, if $s_\theta \in \mathscr{H}(m)$ for all θ.

(ii) $p \in \mathscr{D}$, if $s_{\theta+1} \in \mathscr{D}(m)$, and all preceding terms are in $\mathscr{H}(m)$.

(iii) $p \in \mathscr{B}$, if $s_{\theta+1} \in \mathscr{B}(m)$, and all preceding terms are in $\mathscr{H}(m)$.

Notice that this partition of $\mathscr{C}$ into $\mathscr{H}$, $\mathscr{D}$, and $\mathscr{B}$ is the most obvious adaptation of the partition, defined in the last chapter, of $\mathscr{C}(m)$ into $\mathscr{H}(m)$, $\mathscr{D}(m)$, and $\mathscr{B}(m)$.

The *homology part* of p is

$$s_1, s_2, \ldots, s_\theta, t,$$

where θ is the largest subscript such that s_θ and all preceding terms belong to $\mathscr{H}(m)$, and t is the homology part of $s_{\theta+1}$, as defined in Section II.3 for the monomial language. t may be empty.

The *boundary function* Ω is defined as zero on all elements of $\mathscr{H} \cup \mathscr{B}$, and for all $(s_1, s_2, \ldots)$ in $\mathscr{D}$, written in conventional form,

$$\Omega(s_1, s_2, \ldots) = (s_1, s_2, \ldots, s_\theta, \Omega_m(s_{\theta+1}), s_{\theta+2}, \ldots),$$

where θ is the largest subscript such that s_θ and all preceding terms are in $\mathscr{H}(m)$.

It is evident that Ω extends uniquely to a linear transformation of C, such that (C, Ω) constitutes a chain complex, having ($\mathscr{H}, \mathscr{D}, \mathscr{B}$) as a standard basis.

It will be seen that this complex is isomorphic to (C, Δ), whenever (C, Δ) is free. To have isomorphism in general, it is necessary to introduce a non-zero integer $\tau(p)$ for each p in $\mathscr{D}$ and replace Ω by Ω', which is defined by

$$\Omega'(p) = \tau(p)\Omega(p),$$

where $\tau(p)$ is the coefficient of $\Omega(p)$, which appears in the expansion of $\Delta(p)$ into a linear combination of elements of $\mathscr{C}$.

For instance, in the language of factorizations of the sixth degree monomial $a^2 b^2 c^2$, if

$$p = (a \times b \times c, a \times b \times c),$$

then $\tau(p)$ is 2, because

$$\Delta(p) = -2(a, b \times c, a \times b \times c) + 2(a \times b, c, a \times b \times c) +$$
$$+ 2(ab \times c, a \times b \times c) - 2(a, bc, a \times b \times c)$$

in which $\tau(p)$ is the coefficient of the third term. τ is different from ± 1, only when there are repeated terms of odd degree greater than or equal to 3, since repeated terms of even degree are not allowed in $\mathscr{C}$.

We shall confine our attention to the interpretation Ω, rather than Ω', even though (C, Ω) is not isomorphic to (C, Δ), because Ω is more useful from a programming viewpoint. (C, Ω) is free, which is not true, in general, of (C, Ω') and (C, Δ). Therefore, a program in (C, Ω), which lists the elements of $\mathscr{C}$, requires an input which is generally smaller than the input required for programs in (C, Ω') and (C, Δ).

III.4. THE INTERPRETATION Γ

A chain complex (C, Γ) is now constructed, which is isomorphic to (C, Ω). The importance of it is that it contains a *graphical program*, which lists without repetitions the elements of $\mathscr{C}$, with $\mathscr{H}$ and $\mathscr{D}$, as defined in the last section, being its input and instructions.

$\mathscr{H}$ and $\mathscr{D}$ are defined as in Section III.3. Let p be an element of $\mathscr{D}$ in conventional form,

$$p = (s_1, s_2, \ldots, s_\theta, \ldots),$$

where θ is the largest subscript such that s_θ and all preceding terms are in $\mathscr{H}(m)$. Then, $s_{\theta+1}$ is in $\mathscr{D}(m)$ and is of the form

$$s_{\theta+1} = t \times u,$$

where t is the first factor in $s_{\theta+1}$. The boundary function Γ is defined on $\mathscr{D}$ by the formula

$$\Gamma(p) = -(s_1, s_2, \ldots, s_\theta, t, u, s_{\theta+2}, \ldots) +$$
$$+ \Omega(s_1, s_2, \ldots, s_\theta, \ldots).$$

This definition of Γ is valid, provided that the conventional form is chosen in such a way that the first term

$$(s_1, s_2, \ldots, s_\theta, t, u, s_{\theta+2}, \ldots)$$

does not contain any repetitions of u, when u is even-dimensional. It will be recalled that in the Definition III.1 of *conventional form* the set $\mathscr{H}(m) \cup \mathscr{D}(m)$ was allowed to take any fixed linear ordering, but now it becomes necessary to impose a condition on this ordering, in order that Γ may be well-defined. The following condition is satisfactory and will be used in examples:

$j_1 \times j_2 \times \ldots$ precedes $k_1 \times k_2 \times \ldots$ in the linear ordering of $\mathscr{H}(m) \cup \mathscr{D}(m)$, if the degree of the monomial $j_1 j_2 \ldots$ is less than the degree of $k_1 k_2 \ldots$.

Thus, Γ is well-defined on $\mathscr{D}$, and it is easy to see that

$$\mathscr{H} \cup \mathscr{D} \cup \Gamma \mathscr{D}$$

is a basis for C. Hence, the chain complex is completely determined by requiring Γ to take the value zero on all elements of $\mathscr{H} \cup \Gamma \mathscr{D}$.

III.5. LANGUAGE ISOMORPHISMS

There is an obvious chain isomorphism from (C, Ω) to (C, Γ). It is the identity map, when it is restricted to $\mathscr{H} \cup \mathscr{D}$, and it maps each element $\Omega(d)$ of $\mathscr{B}$ to $\Gamma(d)$.

The remainder of this section is devoted to constructing a chain isomorphism

$$\Psi \colon (C, \Omega) \to (C, \Delta)$$

in the case when (C, Δ) is free. This is accomplished by finding a chain isomorphism

$$\Psi \colon (C, \Omega') \to (C, \Delta)$$

for any (C, Δ), and observing that

$$\Omega' = \Omega,$$

when (C, Δ) is free.

The next lemma makes use of the *shuffle product*, given in

Definition II.1 of Section II.5. Such products will now appear as terms in elements of C. Let

$$s = p \wedge q = \sum \kappa(k_1 \times \ldots \times k_\pi) \in C(m)$$

be a typical product. Then, we adopt the following notation:

$$(\ldots, r, s, t, \ldots) = \sum \kappa(\ldots, r, (k_1 \times \ldots \times k_\pi), t, \ldots).$$

LEMMA III.1. *Let*

$$\Psi_m: (C(m), \Omega_m) \to (C(m), \Delta_m)$$

denote the chain isomorphism of Lemma II.1 for monomial complexes; then, if (C, Δ) is free, a chain isomorphism

$$\Psi: (C, \Omega) \to (C, \Delta)$$

which maps a standard basis $(\mathcal{H}, \mathcal{D}, \mathcal{B})$ into a standard basis $(\mathcal{H}', \mathcal{D}', \mathcal{B}')$ is constructed as follows: Let any element p of $\mathcal{C}$, written in conventional form, be denoted by

$$(s_1, s_2, \ldots, s_\theta, \ldots, s_\sigma),$$

where θ is the largest subscript such that s_θ and all preceding terms are in $\mathcal{H}(m)$, then

$$\Psi(p) = (\Psi_m(s_1), \ldots, \Psi_m(s_{\theta+1}), s_{\theta+2}, \ldots, s_\sigma).$$

Proof. This proof is essentially a repetition of that of Lemma II.1. A linear ordering of $\mathcal{C}$ is found, such that for every p in $\mathcal{C}$, if $\Psi(p)$ is expressed as a linear combination of elements of $\mathcal{C}$ with non-zero coefficients, then p is the largest (with respect to the linear order) of these elements, and its coefficient is $+1$ or -1. This condition is equivalent to saying that Ψ is a change of basis, which is expressible by a square matrix with units (± 1) on the diagonal and zeros above it. If (C, Δ) were not free, we would get numbers on the diagonal with absolute values greater than 1.

Let $\mathscr{C}$ be linearly ordered such that

$$s_1, s_2, \ldots, s_\sigma,$$

written in conventional form, is less than

$$t_1, t_2, \ldots, t_\tau,$$

also written in conventional form, if

(i) $\sigma > \tau$, or

(ii) $\sigma = \tau$, and they are in lexicographic order, relative to that linear ordering of $\mathscr{C}(m)$, which was used in defining the conventional form of elements of $\mathscr{C}$.

For any p in $\mathscr{C}$ it suffices to prove that p is greater than every element q of $\mathscr{C}$, which appears as a term in the expansion of $\Psi(p)$, even though q is not in conventional form. An element of $\mathscr{C}$ can only become smaller (lexicographically) when it is reordered, so as to be in conventional form.

If $p = s_1, s_2, \ldots$, then p is greater than every term in

$$\Psi_m(s_1), \Psi_m(s_2), \ldots,$$

when it is expanded without reordering its factors, because s_ϕ is greater than $\Psi_m(s_\phi)$ for all ϕ by virtue of Lemma II.1.

Suppose there exists p such that it appears in $\Psi(p)$ with a coefficient different from $+1$ or -1. Then, p must contain repeated factors

$$p = (\ldots, s, s, \ldots)$$

which are consecutive, when p is in conventional form, and of odd degree. The coefficient of p in $\Psi(p)$ will equal in absolute value the number of repetitions of the factor s in p. If this is different from 1, then (C, Δ) is not free. Therefore, by assuming that the complex is free, we are certain that Ψ is an isomorphism.

This proves the lemma, and also it proves the following theorem.

THEOREM III.1. *A standard basis* $(\mathscr{H}', \mathscr{D}', \mathscr{B}')$ *of free factoriz-ation complex* (C, Δ) *is as follows: Let the first degree monomials have a linear ordering assigned to them; then*
  (i) $\mathscr{H}'$ *is the set consisting of every element*

$$s_1, s_2, \ldots, s_\phi, \ldots$$

of C, *in which each factor* s_ϕ *is of the form*

$$j_1 \wedge j_2 \wedge \ldots \wedge j_\pi,$$

where $j_1, j_2, \ldots, j_\pi$ *is a properly decreasing sequence of first degree monomials.*
  (ii) $\mathscr{D}'$ *is the set consisting of every element*

$$s_1, s_2, \ldots, s_\theta, s_{\theta+1}, \ldots$$

of C, *in which* s_ϕ *is of the form given in (i), when* $\phi \leqslant \theta$, *and* $s_{\theta+1}$ *is of the form*

$$j_1 \wedge j_2 \wedge \ldots \wedge j_{\lambda-1} \wedge (j_\lambda \times j_{\lambda+1} \times \ldots \times j_\pi),$$

such that $j_1, j_2, \ldots, j_\lambda$ *is a properly decreasing sequence of first* *degree monomials and $j_{\lambda+1}$ is a monomial whose first degree factors are all greater than or equal to j_λ ($1 \leqslant \lambda < \pi$).*

(iii) $\mathscr{B}'$ *is the set of all $\Delta(d)$, such that d is an element of $\mathscr{B}'$.*

III.6. A PROGRAM FOR FACTORIZATIONS

The groups H and D are now completely determined for each of the complexes (C, Δ), (C, Ω), and (C, Γ). Therefore, we have determined the set of all programs in each complex, which list a basis for C.

In particular, $\mathscr{H}$ and $\mathscr{D}$ are the input and instructions, respectively, of a graphical program in (C, Γ), which lists without repetitions the elements of $\mathscr{C}$. In order to determine this program completely, it is still necessary to select a suitable linear ordering for the instructions. It is sufficient to use any linear ordering such that

$$s_1, s_2, \ldots, s_\sigma \text{ precedes } t_1, t_2, \ldots, t_\tau$$

whenever $\sigma \geqslant \tau$. Therefore, one way to specify the program is to give $\mathscr{H}$ in the form of a list without repetitions, and to give $\mathscr{D}$ in the form of an increasing sequence relative to the above type of linear ordering.

A more useful way to specify the program is by induction, as described in Section I.3. This means finding a subroutine $\mathscr{L}$, which assigns to each element x of $\mathscr{C}$ the list $\mathscr{L}(x)$ of all the instructions which depend on x. Then the program is specified by substituting this $\mathscr{L}$ into flowchart 2 of the appendix. It is easy to see from the definition of Γ in Section III.4 how the subroutine $\mathscr{L}$ is to be constructed. For any x in $\mathscr{C}$ it is necessary to list the set $\mathscr{L}(x)$ of all p in $\mathscr{C}$, such that x will be the negative term of $\Gamma(p)$. A subroutine to do this is given by part L of flowchart 6 in the appendix. An entire program in (C, Γ), for listing the elements of $\mathscr{C}$, is specified by flowcharts 2 and 6 of the appendix, taken together. The part of this program which merely lists the set $\mathscr{C}_1$ of factorizations (not the whole language) is specified by flowcharts 1 and 5 of the appendix, taken together.

The programs of this chapter in (C, Ω) and (C, Γ) all have the same number η of input elements. Programs in (C, Δ) also have η input elements, if we admit programs which list multiples of basis elements of C, in those cases when (C, Δ) is not free. η is the smallest possible number of input elements for programs in these languages. It is still a minimum, when we are allowed to use any instruction whatever, which takes any factor

$$j_1 \times j_2 \times \ldots \times j_\pi$$

in any element of $\mathscr{C}$ and replaces it by

$$j_1 \times j_2 \times \ldots \times j_\phi j_{\phi+1} \times \ldots \times j_\pi$$

for any ϕ, depending on the element of $\mathscr{C}$, so as to form a new element of $\mathscr{C}$. This combinatorial result is proved in the next theorem, showing that we have not hampered our languages by the special ways in which we have limited the selection of instructions.

THEOREM III.2. *Let $\mathscr{C}$ be the factorization language associated with a given monomial m, and let η be the number of factorizations of m into homological form. Let $\mathscr{H}, \mathscr{D}, \mathscr{B}$ be any partition of the set $\mathscr{C}$ into disjoint subsets, such that there exists a one-to-one correspondence Θ of $\mathscr{D}$ onto $\mathscr{B}$ of the form*

$$\Theta(\ldots, (j_1 \times j_2 \times \ldots \times j_\pi), \ldots)$$
$$= (\ldots, (j_1 \times j_2 \times \ldots \times j_\phi j_{\phi+1} \times \ldots \times j_\pi), \ldots)$$

where ϕ varies with the elements of $\mathscr{D}$ and

$$1 \leqslant \phi < \pi.$$

Under these conditions, if Θ is chosen such that $\mathscr{D}$ and $\mathscr{B}$ are as large as possible, then $\mathscr{H}$ has η elements.

Proof. Let $\mathscr{C}$ be partitioned into a family of subsets, where each subset $\mathscr{C}(p)$ corresponds to a single element p of $\mathscr{C}_1$:

$$\mathscr{C}(p) =$$
$$= \{(j_{11} \times j_{12} \times \ldots, j_{21} \times j_{22} \times \ldots, j_{31} \times j_{32} \times \ldots, \ldots):$$
$$(j_{11}j_{12} \ldots, j_{21}j_{22} \ldots, j_{31}j_{32} \ldots, \ldots) = p\}.$$

Since p and $\Theta(p)$ are always in the same subset, the number of elements which are not in any of the pairs, defined by Θ, is given by

$$\sum_p \eta(p),$$

where $\eta(p)$ is the number of unpaired elements in $\mathscr{C}(p)$.

In any finite set the absolute minimum of unpairable elements is 0 or 1, according as the number of elements in the set is even or odd. This minimum is attained by the pairing Ω, applied to $\mathscr{C}(p)$, because this set contains at most one element h in $\mathscr{H}$. This element h exists if all the terms of p are square-free, in which case h is of the form

$$(j_{11} \times j_{12} \times \ldots, j_{21} \times j_{22} \times \ldots, j_{31} \times j_{32} \times \ldots, \ldots)$$

where $(j_{11}, j_{12}, \ldots), (j_{21}, j_{22}, \ldots), (j_{31}, j_{32}, \ldots), \ldots$ are decreasing sequences of first degree monomials. Hence η equals the number of elements not paired up by Ω, which proves the theorem.

	$\mathscr{C}_4$	$\mathscr{C}_3$	$\mathscr{C}_2$	$\mathscr{C}_1$
$\mathscr{H}_1$	=			a, a, b, b
$\mathscr{L}(a, a, b, b)$	=		$\begin{cases} b, b, a \times a \to b, b, a^2 \\ a, b, a \times b \to a, b, ab \\ a, a, b \times b \to a, a, b^2 \end{cases}$	
$\mathscr{L}(a, b, ab)$	=		$\begin{cases} b, a \times ab \to b, a^2 b \\ a \times b, ab \to ab, ab \end{cases}$	
$\mathscr{L}(a, a, b^2)$	=		$\begin{cases} a \times a, b^2 \to a^2, b^2 \\ a, a \times b^2 \to a, ab^2 \end{cases}$	
$\mathscr{L}(a, ab^2)$	=		$a \times ab^2 \to a^2 b^2$	
$\mathscr{H}_2$	=		$a, b, b \times a$	
$\mathscr{L}(a, b, a \times b)$	=	$b, a \times a \times b \to b, a^2 \times b$		
$\mathscr{L}(a, a, b \times b)$	=	$\begin{cases} a \times a, b \times b \to a^2, b \times b \\ a, a \times b \times b \to a, ab \times b \\ a, b \times a \times b \to a, b \times ab \end{cases}$		
$\mathscr{L}(a, a \times b^2)$	=	$a \times a \times b^2 \to a^2 \times b^2$		
$\mathscr{L}(a, b, b \times a)$	=	$\begin{cases} a \times b, b \times a \to ab, b \times a \\ b, a \times b \times a \to b, ab \times a \\ b, b \times a \times a \to b, b \times a^2 \\ a, b \times b \times a \to a, b^2 \times a \end{cases}$		
$\mathscr{L}(a, ab \times b)$	=	$a \times ab \times a \to a^2 b \times b$		
$\mathscr{L}(a, b \times ab)$	=	$\begin{cases} a \times b \times ab \to ab \times ab \\ b \times a \times ab \to b \times a^2 b \end{cases}$		
$\mathscr{L}(b, b \times a^2)$	=	$b \times b \times a^2 \to b^2 \times a^2$		
$\mathscr{L}(a, b^2 \times a)$	=	$a \times b^2 \times a \to ab^2 \times a$		
$\mathscr{L}(a, a \times b \times b) =$	$a \times a \times b \times b \to a^2 \times b \times b$			
$\mathscr{L}(a, b \times a \times b) =$	$\begin{cases} a \times b \times a \times b \to ab \times a \times b \\ b \times a \times a \times b \to b \times a^2 \times b \end{cases}$			
$\mathscr{L}(b, b \times a \times a) =$	$b \times b \times a \times a \to b^2 \times a \times a$			
$\mathscr{L}(a, b \times b \times a) =$	$\begin{cases} a \times b \times b \times a \to ab \times b \times a \\ b \times a \times b \times a \to b \times ab \times a \end{cases}$			

III.7. SAMPLE PROGRAMS

To demonstrate the kind of program developed in this chapter, let us list all the elements of the three factorization languages, which are associated with the monomials a^2b^2, a^5, and abc, respectively.

It is necessary to agree upon the conventional form of the language elements. Each element is a list, whose entries are taken from the monomial language $\mathscr{C}(m)$, where m is a^2b^2, a^5, or abc, and it is in conventional form when its entries are ordered so as to conform with a fixed linear ordering of $\mathscr{C}(m)$. By virtue of Definition III.1 it is only necessary to fix the linear ordering of the subset $\mathscr{H}(m) \cup \mathscr{D}(m)$ of $\mathscr{C}(m)$. This will be done exactly as it was in the proof of Lemma II.1. That is, $j_1 \times j_2 \times \ldots$ will precede $k_1 \times k_2 \times \ldots$, if the degree of the monomial $j_1 j_2 \ldots$ is less than that of $k_1 k_2 \ldots$, and, if these degrees are equal, then the order of precedence is lexicographic.

EXAMPLE III.1. Use a program in (C, Γ) to list all the elements of the factorization language associated with the monomial a^2b^2. The input $\mathscr{H}$ has two entries, (a, a, b, b) and $(a, b, b \times a)$. We follow flowchart 2 of the appendix. Each line in the format above represents one passage through the inner loop of the flowchart, but no line is written when $\lambda = 0$. The lines are indented in such a way that the elements of $\mathscr{C}_1$, $\mathscr{C}_2$, $\mathscr{C}_3$, and $\mathscr{C}_4$ appear in separate columns, one for each passage through the outer loop. An arrow $\rightarrow$ runs from each instruction to the element it produces.

EXAMPLE III.2. For any integer ν the factorization language associated with the monomial a^ν may be described as the *language of partitions of the integer* ν, because the factorizations of a^ν coincide with the partitions of ν. Since there is only one indeterminate in this language, every monomial a^λ can be expressed simply by the integer λ.

The following is a table/diagram (printed sideways on the page) relating the objects $\mathscr{C}_5,\ \mathscr{C}_4,\ \mathscr{C}_3,\ \mathscr{C}_2,\ \mathscr{C}_1$ through refinement arrows ($\to$).

Object	$\mathscr{C}_5$	$\mathscr{C}_4$	$\mathscr{C}_3$	$\mathscr{C}_2$	$\mathscr{C}_1$
$\mathscr{H}_1$	=				$1,1,1,1,1$
$\mathscr{L}(1,1,1,1,1)$	=			$1,1,1,1\times1 \to$	$1,1,1,1,2$
$\mathscr{L}(1,1,1,2)$	=			$\left\{\begin{array}{l}1,1,1\times1,2\\ 1,1,1\times1,2\end{array}\right. \to$	$\left\{\begin{array}{l}1,2,2\\ 1,1,3\end{array}\right.$
$\mathscr{L}(1,1,3)$	=			$\begin{array}{l}2,1\times2 \to\\ 1,2\times2 \to\end{array}$	
$\mathscr{L}(1,4)$	=			$2\times3 \to$	$\left\{\begin{array}{l}2,3\\ 1,4\end{array}\right.$
$\mathscr{L}(1,1,1\times1)$	=		$\left\{\begin{array}{l}1,1,1\times1\times1\\ 1\times1,1\times2\end{array}\right. \to$	$\begin{array}{l}2,2\times1 \to\\ 1,3\times1 \to\end{array}$	5
$\mathscr{L}(1,1,1\times2)$	=		$\begin{array}{l}1,1\times1\times2 \to\\ 1,1\times1\times2 \to\end{array}$	$3\times2 \to$	
$\mathscr{L}(1,1\times3)$	=		$1\times1\times3 \to$	$4\times1 \to$	
$\mathscr{L}(1,1,2\times1)$	=		$\left\{\begin{array}{l}1\times1,2\times1\\ 1,1\times2\times1\end{array}\right. \to$	$2,2\times1$	
$\mathscr{L}(1,2\times2)$	=		$1\times2\times2 \to$		
$\mathscr{L}(1,3\times1)$	=		$1\times3\times1 \to$		
$\mathscr{L}(1,1,1\times1\times1)$	=	$\left\{\begin{array}{l}1\times1,1\times1\times1\\ 1,1\times1\times1\times1\end{array}\right. \to$	$\begin{array}{l}2,1\times1\times1 \to\\ 1,2\times1\times1 \to\end{array}$		
$\mathscr{L}(1,1\times1\times2)$	=	$1\times1\times1\times2 \to$	$2\times1\times1\times2 \to$		
$\mathscr{L}(1,1\times2\times1)$	=	$1\times1\times2\times1 \to$	$2\times2\times1 \to$		
$\mathscr{L}(1,2\times1\times1)$	=	$1\times2\times1\times1 \to$	$2\times2\times1 \to$		
$\mathscr{L}(1,1\times1\times1\times1) = 1\times1\times1\times1\times1 \to 2\times1\times1\times1$	=	$2\times1\times1\times1 \to$	$3\times1\times1$		

Use a program in (C, Γ) to list all the elements in the language of partitions of the integer 5. The input $\mathscr{H}$ contains only the partition

1, 1, 1, 1, 1

of 5 into ones. The format, used above, is as described in Example III.1.

EXAMPLE III.3. The factorization language associated with a square-free monomial may be described as the *language of partitions of a set*, because the factorizations reduce to partitions of the set of indeterminates.

Use a program in (C, Γ) to list all the elements in the language of partitions of the set $\{a, b, c\}$. We use the notation of the language of factorizations of abc. The input $\mathscr{H}$ consists of the following elements:

a, b, c

$a, c \times b$

$b, c \times a$

$c, b \times a$

$c \times b \times a.$

The format, used below, is as described in Example III.1.

		$\mathscr{C}_3$	$\mathscr{C}_2$	$\mathscr{C}_1$
$\mathscr{H}_1$	$=$			a, b, c
$\mathscr{L}(a, b, c)$	$=$		$\begin{cases} c, a \times b \to c, ab \\ b, a \times c \to b, ac \\ a, b \times c \to a, bc \end{cases}$	
$\mathscr{L}(a, bc)$	$=$		$a \times bc \to abc$	
$\mathscr{H}_2$	$=$		$\begin{cases} a, c \times b \\ b, c \times a \\ c, b \times a \end{cases}$	
$\mathscr{L}(a, bc)$	$=$	$\begin{cases} a \times b \times c \to \\ b \times a \times c \to \end{cases}$	$\begin{array}{l} ab \times c \\ b \times ac \end{array}$	
$\mathscr{L}(a, cb)$	$=$	$\begin{cases} a \times c \times b \to \\ c \times a \times b \to \end{cases}$	$\begin{array}{l} ac \times b \\ c \times ab \end{array}$	
$\mathscr{L}(b, ca)$	$=$	$b \times c \times a \to$	$bc \times a$	
$\mathscr{H}_3$	$=$	$c \times b \times a$		

Chapter IV

Transpositions

IV.1. THE LANGUAGE

This chapter is concerned with how to make a list of all permutations of a given finite sequence s by means of transposition instructions. This means to list without repetitions every element of the set $\mathscr{C}_1$ of all sequences which can be made by permuting the terms of s, and to do so by means of a program, whose instructions have to be taken from the set $\mathscr{C}_2$ of transpositions. Each element of $\mathscr{C}_2$ depends on a particular element of $\mathscr{C}_1$, which it converts into a new one by transposing two adjacent terms. For instance, if

$$s = (a, a, b, b),$$

then there is an instruction in $\mathscr{C}_2$, denoted by $b, a, b \times a$, which produces
$$(b, a, b, a) \in \mathscr{C}_1$$

by transposing the third and fourth terms of

$$(b, a, a, b) \in \mathscr{C}_1.$$

Using the theory developed in Chapter I, we embed $\mathscr{C}_1$ and $\mathscr{C}_2$ in a programming language and examine the set of all programs in the language, which list every element of the language. Programs which list just the elements of $\mathscr{C}_1$ or $\mathscr{C}_2$ are special cases. Finally, by way of application, we select and construct a program which lists the whole language. That part of the program which lists the elements of $\mathscr{C}_1$ is fairly simple and might have been discovered by other means. It can be described as follows:

The first entry on the list is a finite sequence, whose terms are in non-decreasing order relative to an arbitrarily chosen linear ordering. Then the rule for completing the list of all permutations of this sequence is to look at each sequence t already on the list and adjoin to the list every sequence t' obtainable from t by transposing an adjacent pair of its terms, provided that these two terms become the first adjacent pair in t' which are in decreasing order. For instance, suppose the first entry is

$$s_1 = a, a, b, b.$$

Then, there is one permissible transposition, by means of which s_1 becomes

$$s_2 = a, b, a, b.$$

s_2 yields two entries:

$$s_3 = b, a, a, b$$

$$s_4 = a, b, b, a.$$

s_3 yields nothing, and s_4 yields

$$s_5 = b, a, b, a.$$

s_5 yields

$$s_6 = b, b, a, a.$$

s_6 yields nothing, and the list is complete, there being no further entries to apply transpositions to.

Now we are prepared to make a formal construction of a language of transpositions. Its elements will be given various interpretations in the sections which follow, but essentially it consists of a set $\mathscr{C}_1$ of sequences, a set $\mathscr{C}_2$ of transposition instructions, which generate elements of $\mathscr{C}_1$, and for each $\pi > 2$ a set $\mathscr{C}_\pi$ of generalized transposition instructions which generate elements of $\mathscr{C}_{\pi-1}$.

The *language of transpositions* is of the form

$$\mathscr{C} = \mathscr{C}_1 \cup \mathscr{C}_2 \cup \ldots \cup \mathscr{C}_\pi \cup \ldots \cup \mathscr{C}_\rho,$$

where

(i) $\mathscr{C}_1$ is the set of all sequences with terms from the set

$$\mathscr{M} = \{a_1, a_2, \ldots, a_\mu\},$$

such that each a_θ appears exactly ρ_θ times in the sequence. The length of each sequence is ρ, where

$$\rho = \rho_1 + \rho_2 + \ldots + \rho_\mu.$$

(ii) $\mathscr{C}_2$ is the set of all the $(\rho - 1)$-termed sequences, called transpositions, which can be formed by replacing two adjacent terms $e_\theta, e_{\theta+1}$ in any element of $\mathscr{C}_1$ by a term $e_\theta \times e_{\theta+1}$. This symbolism suggests that e_θ and $e_{\theta+1}$ have changed places.

(iii) $\mathscr{C}_\pi$ is the set consisting of every expression which can be obtained from an element $\mathscr{C}_1$ by putting the symbol $\times$ between adjacent terms in exactly $(\pi - 1)$ positions. An element of this set is said to be of *dimension* π, and it is expressed by a sequence

$$(l_1, l_2, \ldots, l_\theta, \ldots, l_{\rho-\pi+1}),$$

where each l_θ is of the form,

$$l_\theta = e_1 \times e_2 \times e_3 \times \dots ,$$

in which $e_1, e_2, \dots$ are elements of the set $\mathscr{M}$, where

$$\mathscr{M} = \{a_1, a_2, \dots, a_\mu\}.$$

Notice that this language contains elements such as $e \times e$, which have consecutive pairs of equal terms within a block. It simplifies the definition of the language to include such elements, even though they belong to that trivial kind of instruction which transposes identical symbols.

IV.2. THE INTERPRETATION Δ

An interpretation Δ of $\mathscr{C}$ is defined by embedding $\mathscr{C}$ in a chain complex (C, Δ), in which C is the free Abelian group generated by $\mathscr{C}$ and Δ is the boundary function, which is determined by giving a value to $\Delta(p)$ for each $p \in \mathscr{C}$. First, we consider the case where each a_ϕ appears exactly once in p for $1 \leqslant \phi \leqslant \mu$, that is

$$1 = \rho_1 = \rho_2 = \dots = \rho_\mu.$$

(i) If $p \in \mathscr{C}_1$, then $\Delta(p) = 0$.
(ii) If $p \in \mathscr{C}_2$. then it is of the form

$$p = (l_1, l_2, \dots, l_{\mu-1}),$$

where each term is of the form a_ϕ except for one l_θ, which is of the form

$$l_\theta = a_\kappa \times a_\lambda,$$

and

$$\Delta(p) = (\ldots, l_{\theta-1}, a_\kappa, a_\lambda, l_{\theta+1}, \ldots) -$$
$$- (\ldots, l_{\theta-1}, a_\lambda, a_\kappa, l_{\theta+1}, \ldots).$$

It is also written as

$$\Delta(p) = (\ldots, l_{\theta-1}, (a_\kappa, a_\lambda - a_\lambda, a_\kappa), l_{\theta+1}, \ldots)$$

or as

$$\Delta(p) = (\ldots, l_{\theta-1}, \Delta(l_\theta), l_{\theta+1}, \ldots).$$

This is a formal way of saying that, when p is in $\mathscr{C}_2$, it is interpreted as a transposition of two adjacent terms a_κ and a_λ in an element of $\mathscr{C}_1$. Next, this interpretation is generalized to cover all elements of $\mathscr{C}$.

(iii) If $p \in \mathscr{C}_\pi$, then it is a sequence of the form

$$p = (l_1, l_2, \ldots, l_\theta, \ldots, l_{\mu-\pi+1}),$$

where each l_θ is again a sequence of the form

$$l_\theta = e_1 \times e_2 \times \ldots \times e_\lambda,$$

and

$$\Delta(p) = \sum_{\theta=1}^{\mu-\pi+1} \epsilon_\theta (l_1, l_2, \ldots, \Delta(l_\theta), \ldots, l_{\mu-\pi+1}),$$

in which ϵ_θ is $+1$ or -1 according as the total number of x's in

$$(l_1, l_2, \ldots, l_{\theta-1})$$

is even or odd, and

$$\Delta(l_\theta) = \sum (-1)^{\kappa-1} \epsilon(f_1 \times \ldots \times f_\kappa, g_1 \times \ldots \times g_{\lambda-\kappa}),$$

in which the sum is taken over every proper subsequence $(f_1, \ldots, f_\kappa)$ of $(e_1, \ldots e_\lambda)$ with $(g_1, \ldots g_{\lambda-\kappa})$ being the subsequence of $(e_1, \ldots, e_\lambda)$ complementary to $(f_1, \ldots f_\kappa)$, and ϵ is $+1$ or -1 according as the sequence

$$(f_1, \ldots, f_\kappa, g_1, \ldots, g_{\lambda-\kappa})$$

is an even or odd permutation of $(e_1, \ldots, e_\lambda)$.

The expression

$$(l_1, l_2, \ldots, \Delta(l_\theta), \ldots, l_{\rho-\pi+1})$$

is expanded linearly, that is, by the rules

$$(\ldots, j, \kappa k, l, \ldots) = \kappa(\ldots, j, k, l, \ldots),$$

where κ is an integer, and

$$(\ldots, j, k + l, m, \ldots) =$$
$$= (\ldots, j, k, m, \ldots) + (\ldots, j, l, m, \ldots).$$

LEMMA IV.1. (C, Δ) is a chain complex.

Proof. It must be shown that

$$\Delta\Delta = 0.$$

What Δ does, essentially, is to separate every term l_θ of p into an ordered pair of disjoint proper subsequences in every possible way. There is a duality between this operation and the shuffle product, defined in Section II.1 That is, l_θ separates into

$$f_1 \times \ldots \times f_\kappa, g_1 \times \ldots \times g_{\lambda-\kappa}$$

if, and only if, l_θ is one of the terms of the shuffle product

$$(f_1 \times \ldots \times f_\kappa) \wedge (g_1 \times \ldots \times g_{\lambda-\kappa}).$$

From this fact it follows that Δ is the dual of a coboundary function

$$\Delta^* : C^* \to C^*$$

defined as follows:

C^* is the free Abelian group generated by every homomorphism p^* of C into the integers such that for any $q \in C$

$$p^*(q) = \begin{cases} 1, \text{ if } q = p. \\ 0, \text{ if } q \neq p. \end{cases}$$

Δ^* has the following formula:

$$\Delta^*(p^*) = \sum_{\theta=2}^{\mu-\pi+1} \epsilon_\theta (l_1, \ldots, l_{\theta-1} \wedge l_\theta, \ldots l_{\mu-\pi+1})^*,$$

where ϵ_θ is $+1$ or -1 according as the number of x's in $l_1, \ldots l_{\theta-1}$ is even or odd.

It is easy to see that

$$\Delta^* \Delta^* = 0,$$

and that Δ^* is dual to Δ, in the sense that for any $x \in C$

$$(\Delta^*(p^*))(x) = p^*(\Delta(x)).$$

Hence,

$$0 = (\Delta^*\Delta^*(p^*))(x) = p^*(\Delta\Delta(x)),$$

and, since p^* was any basis element of C^*,

$$\Delta\Delta(x) = 0$$

for all x. This proves the lemma.

The definition of (C, Δ) can now be extended to the case, in which each element of $\mathscr{C}$ contains $\rho_1, \rho_2, \ldots, \rho_\mu$ appearances of the elements $a_1, a_2, \ldots, a_\mu$, respectively.
Let

$$\rho = \rho_1 + \rho_2 + \ldots + \rho_\mu.$$

The above formula for Δ is applied to each element of p of $\mathscr{C}$ as if the ρ elements appearing in it were different.

It was noticed earlier that elements of $\mathscr{C}$, in which x stands between equal values of a_ϕ, are trivial in $\mathscr{C}_2$. All such elements generate a subcomplex (C^0, Δ) in (C, Δ) which could be factored out, by virtue of the fact that

$$\Delta(C^0) \subset C^0.$$

The quotient complex C/C^0 has, as a basis, the language $\mathscr{C} - \mathscr{C}^0$, obtained from $\mathscr{C}$ by striking out elements which contain

$$\ldots \times a_\phi \times a_\phi \times \ldots.$$

But, in this text we preserve such elements for the sake of generality. For example, the language $\mathscr{C} - \mathscr{C}^0$ associated with the sequence

$$(a, a, \ldots, a)$$

has only one element, but $\mathscr{C}$ has $2^{\mu-1}$, where μ is the number of a's. The algorithms which list $\mathscr{C}$ are of combinatorial interest, even though the sequence we start with is trivial.

IV.3. THE INTERPRETATION Ω

A chain complex (C, Ω), having a standard basis denoted by ($\mathscr{H}$, $\mathscr{D}, \mathscr{B}$) is to be constructed. It will be shown to be isomorphic to (C, Δ) whenever (C, Δ) is free, and otherwise it is isomorphic to the largest free subcomplex of (C, Δ). One case, when (C, Δ) is free, is when

$$1 = \rho_1 = \rho_2 = \ldots = \rho_\mu.$$

Let p be any element of $\mathscr{C}$, the language of transpositions. p can be expressed as

$$p = (l_1, l_2, \ldots, l_\theta, \ldots),$$

in which each l_θ is called a *block* and is of the form

$$l_\theta = e_1 \times e_2 \times \ldots \times e_\lambda,$$

and $e_1, e_2, \ldots, e_\lambda$ are elements of the given set $\mathscr{M}$, where

$$\mathscr{M} = \{a_1, a_2, \ldots, a_\mu\}.$$

The definition of $\Omega(p)$ depends on the fact that there is a unique prime factorization theorem for blocks. Accordingly, we must digress now to define *prime* blocks, where a block is nothing more than a finite sequence written in multiplicative notation, and develop the theory of prime factorization of blocks or sequences.

As a means of establishing an order relation between any two blocks, let us adopt the lexicographic ordering relative to the alphabet

$$a_1 < a_2 < a_3 < \dots,$$

but when one block j is a factor of another block $j \times k$, let us make their ordering depend on that of j and k, rather than use the dictionary method, which always puts j before $j \times k$. When two blocks contain the same number of factors, as, for instance, $j \times k$ and $k \times j$, then there is no doubt about what is meant by the lexicographic ordering. In the general case we adopt the following definition (*cf.* reference 2):

DEFINITION IV.1. The *natural lexicographic* ordering of two blocks j and k, whose factors belong to a given linearly ordered set $(\mathscr{M}, <)$, is defined as follows: $j < k$ if $j \times k$ precedes $k \times j$ in the common lexicographic ordering (with $\mathscr{M}$ as alphabet), and $j \simeq k$, if $j \times k = k \times j$.

In the above definition $j \simeq k$ defines an equivalence relation, and $j < k$ determines a linear ordering of the equivalence classes. It is easy to see that equivalence of two blocks means they are both powers of the same block l, and, if l is chosen to be as small as possible, then the equivalence class equals

$$\{l, l \times l, l \times l \times l, \dots\}$$

LEMMA IV.2. *If $j < k$, then $j < j \times k < k \times j < k$, and if, $j \simeq k$, then $j \simeq j \times k = k \times j \simeq k$.*

Proof. Let $[j, k]$ be defined as taking one of the values $<$, $\simeq$, or $>$, whichever holds between j and k, and let us verify that

$$[j, k] = [j, j \times k] = [j, k \times j]$$

which is another way to state the lemma.

$$[j, k] = [j \times k, k \times j] = [j \times j \times k, j \times k \times j] = [j, j \times k]$$

$$[j, k] = [j \times k, k \times j] = [j \times k \times j, k \times j \times j] = [j, k \times j].$$

In each of the above two rows the middle equality holds, because blocks of equal length are involved. The other equal signs stem directly from Definition IV.1. This proves the lemma, including the statement, made earlier, that j and $j \times k$ are ordered in the same manner as j and k.

DEFINITION IV.2. A block is *prime* if $j < k$ for every factorization $j \times k$ of the block.

LEMMA IV.3. *Suppose a block has a factorization $j \times k$, such that $j > k$ or $j \simeq k$, then j is prime, if it is the smallest such factor.*

Proof. Let $j_1 \times j_2$ be any factorization of j. Therefore,

$$j_1 < j_2 \times k \text{ and } j_1 \times j_2 \gtrless k.$$

or

$$j_1 < j_1 \times j_2 \times k \text{ and } j_1 \times j_2 \times k \lessgtr j_1 \times j_2$$

Then, by the transitivity of the ordering, which is easily shown,

$$j_1 < j_1 \times j_2 \text{ or } j_1 < j_2.$$

This proves that j is prime.

LEMMA IV.4. *Let j be a prime block; then $j < k$ implies $j < kl$ for any blocks k and l.*

Proof. There are three cases:
(i) j is not a left-factor of k, and k is not a left-factor of j. (x is a

left-factor of y, when there exists z such that $x \times z = y$; or in the language of sequences x is a *partial sequence* of y.) In this case the lemma is obviously true.

(ii) k is a left-factor of j. This case never applies, because, if

$$j = k \times i,$$

then $k \times i < k$ and $i < k$, which is impossible when j is prime.

(iii) j is a left-factor of k. Hence, $j \times i = k$, and what we wish to prove is

$$j < j \times i \text{ implies } j < j \times i \times l.$$

This is equivalent to

$$j < i \text{ implies } j < i \times l.$$

Thus our lemma is reduced to a statement in which k is replaced by i. By a repeated application of such reduction, the lemma will eventually reach case (i), the verification of which is easy.

THEOREM IV.1. *Any block (finite sequence) is uniquely express-ible as a product*

$$l_1 \times l_2 \times l_3 \times \ldots$$

of prime blocks (sequences), such that

$$l_1 \geq l_2 \geq l_3 \geq \ldots$$

in the natural lexicographic ordering.

Proof. Start with block s, and let l_1 be the shortest block with

$$s = l_1 \times s_1 \text{ and } l_1 \gtrless s_1.$$

Let l_2 be the shortest block with

$$s_1 = l_2 \times s_2 \text{ and } l_2 \gtrless s_2.$$

Continue thus until s has been reduced to

$$s = l_1 \times l_2 \times l_3 \times \ldots$$

All the l's are prime by Lemma IV.3.
  In general,

$$s_{\nu-1} = l_\nu \times s_\nu = l_\nu \times l_{\nu+1} \times s_{\nu+1} \text{ and } l_\nu \gtrless l_{\nu+1} \times s_{\nu+1}.$$

It follows from Lemma IV.4 that

$$l_\nu \gtrless l_{\nu+1}$$

and, since the l's cannot be prime powers, that

$$l_\nu \gtrless l_{\nu+1}.$$

It remains to be shown that the factorization of s is unique. Assume there are two distinct factorizations,

$$s = k_1 \times k_2 \times k_3 \times \ldots = l_1 \times l_2 \times l_3 \times \ldots ,$$

which satisfy the lemma. Then there is a smallest θ such that $k_\theta \neq l_\theta$, one of which must be a left factor of the other. If l_θ is a left-factor of k_θ, let us write $l_\theta \subset k_\theta$. There exists $\phi > \theta$

$$l_\theta \times l_{\theta+1} \times \ldots \times l_\phi \subset k_\theta,$$

while

$$l_\theta \times l_{\theta+1} \times \ldots \times l_\phi \times l_{\phi+1} \supset k_\theta.$$

Obviously, neither of these can reduce to equality.

Define t by

$$l_\theta \times l_{\theta+1} \times \ldots \times l_\phi \times t = k_\theta$$

and, since k_θ is prime

$$l_\theta \times l_{\theta+1} \times \ldots \times l_\phi < t.$$

However, this is contradicted by the following application of Lemma IV.2:

$$l_\theta \gtrless l_{\theta+1}$$

$$l_\theta \times l_{\theta+1} \gtrless l_{\theta+1} \gtrless l_{\theta+2}$$

$$l_\theta \times l_{\theta+1} \times l_{\theta+2} \gtrless l_{\theta+2} \gtrless l_{\theta+3}$$

continuing thus, we obtain

$$l_\theta \times l_{\theta+1} \times \ldots \times l_\phi \gtrless l_\phi \gtrless l_{\phi+1} > t,$$

in which the last inequality follows from the fact that $l_{\phi+1}$ is prime. This is a contradiction, which shows that distinct factorizations could not have existed. This proves the theorem.

Now, we may return to the construction of the chain complex (C, Ω), which will be done by giving the value of $\Omega(p)$ for any p in $\mathscr{C}$. p is a sequence of blocks, which has been expressed by

$$p = (l_1, l_2, \ldots, l_\theta, \ldots),$$

and each block l_θ is uniquely expressible as a non-increasing sequence of *prime factors*.

$$l_\theta = f_1 \times f_2 \times \ldots \times f_\phi.$$

Accordingly, p is uniquely expressible as a sequence of primes, where some consecutive pairs are separated by a comma and others by the symbol $\times$. p is in *homological form* if the following two conditions hold:

(i) Every block l_θ in p is either prime, or it is of the form $l \times l$, where l is a prime block of odd dimension.

(ii) For every consecutive pair l_θ, $l_{\theta+1}$ of blocks, $l_\theta < l_{\theta+1}$ or $l_\theta \simeq l_{\theta+1}$, but in an interval of two or more equivalent blocks only the first can be a product $l \times l$ of prime blocks of odd dimension.

In other words, there is only one circumstance, which prevents us from defining p as a non-decreasing sequence of prime blocks, which is the possibility of a block of the form $l \times l$. Such a block cannot be preceded by l, it cannot be repeated, and l cannot be even-dimensional.

If p is not in homological form, then it is possible to separate it at a position between two of its prime factors, such that the part on the left of it is in homological form. This separation can occur at a comma or at a symbol $\times$, and its position, is uniquely determined by choosing it in the following manner:

p is separated into a *homological part q* on the left and a remaining part r on the right, such that

$$p = (q, r) \text{ or } p = q \times r,$$

by first reducing every block in p to its non-increasing sequence of prime factors, and then finding the first separation between consecutive prime factors, which is of one of the following types. (These are the ways in which p can fail to be in homological form).

(i) The separation may fall between consecutive prime factors f and g, if $f > g$.

(ii) If f is even-dimensional, the separation may fall between consecutive prime factors f and g, if $f = g$.

(iii) If f is odd-dimensional, the separation may fall between f or $f \times f$ on the left of it and an even number of f's to the right of it in the form $f \times f \times \ldots \times f$, where each f is a prime factor of p.

This method of finding the homological part of p is accomplished by examining p from left to right, and the prime factorization of p is also carried out from left to right, as shown in the proof of Theorem IV.1. Therefore, in order to determine the homological part of p, it is only necessary to factorize it up to the point at which it ceases to be in homological form.

It should be emphasized that the factor f used above is always a prime factor of p, as defined in the unique factorization theorem (Theorem IV.1). Suppose

$$j = f \times f \times f \times g,$$

where f is prime and odd-dimensional, g is prime, and $f < g$. Then the separation, defined in (iii) does not apply, because j is prime and, therefore, there is no position at which it can be separated. Prime factorization relative to a non-commutative product is not independent of the ordering of the factors. Accordingly, $f \times f \times f \times g$ is prime according to Definition IV.2, while $g \times f \times f \times f$, for example, reduces to four prime factors.

Let us now use the separation of p into a homological part q and a remaining part r to define a partition of the basis $\mathscr{C}$ of C into the three parts $\mathscr{H}$, $\mathscr{D}$, and $\mathscr{B}$ which will constitute a standard basis for (C, Ω).

(i) $p \in \mathscr{H}$ if p is in homological form. Consequently, when p is not in $\mathscr{H}$ there exists a separation of p into $q \times r$ or (q, r), where q is the homological part.

(ii) $p \in \mathscr{D}$ if the separation of p occurs in the middle of a block, so that

$$p = q \times r.$$

(iii) $p \in \mathscr{B}$ if the separation of p occurs between two consecutive blocks, so that

$$p = q, r.$$

The *boundary function* Ω is defined as taking the value zero on all elements of $\mathscr{H} \cup \mathscr{B}$. If $p \in \mathscr{D}$, then

$$p = q \times r,$$

where q is its homological part, and

$$\Omega(p) = (q, r).$$

Thus, Ω is defined on a basis of C and extends uniquely to a linear transformation of C into itself. It is evident that (C, Ω) is a free chain complex, having $(\mathscr{H}, \mathscr{D}, \mathscr{B})$ as a standard basis.

It will be proved later that (C, Ω) is isomorphic to (C, Δ), whenever the latter is free. To have had isomorphism in the general case of any (C, Δ), it would have been necessary to introduce a non-zero integer $\tau(p)$ for each p in $\mathscr{D}$ and to replace Ω by Ω', which is defined by

$$\Omega'(p) = \tau(p)\Omega(p).$$

However, Ω provides an interpretation of $\mathscr{C}$ which, though different, is stronger than the interpretations Δ and Ω' from a programming viewpoint, in the sense that a program in (C, Ω) for listing the elements of $\mathscr{C}$ requires less input than such programs in (C, Ω')

and (C, Δ). To look at it another way, if the input for a program in (C, Ω) were used as input for programs in either of the other two complexes, it would only be possible to list integral multiples of elements of $\mathscr{C}$.

IV.4. THE INTERPRETATION Γ

A chain complex (C, Γ) is now constructed, which is isomorphic to (C, Ω). The importance of it is that it contains a graphical program, which lists without repetitions the elements of $\mathscr{C}$, with $\mathscr{H}$ and $\mathscr{D}$, as defined in the last section, being its input and instructions.

Let p be any element of $\mathscr{D}$,

$$p = l_1, l_2, \ldots, l_\theta, \ldots,$$

whose homological part includes some, but not all, of the block l_θ, where

$$l_\theta = e_1 \times e_2 \times \ldots \times e_\kappa.$$

The boundary function Γ is defined on $\mathscr{D}$ by the formula

$$\Gamma(p) =$$
$$= \Omega(p) - (l_1, \ldots, l_{\theta-1}, e_2 \times \ldots \times e_\kappa, e_1, l_{\theta+1}, \ldots).$$

It is easy to see that

$$\mathscr{H} \cup \mathscr{D} \cup \Gamma \mathscr{D}$$

is a basis for C. Hence, a chain complex (C, Γ) is completely determined by requiring Γ to take the value zero on every element of $\mathscr{H} \cup \Gamma \mathscr{D}$.

IV.5. LANGUAGE ISOMORPHISMS

There is an obvious chain isomorphism from (C, Ω) to (C, Γ). It is the identity map, when restricted to $\mathscr{H} \cup \mathscr{D}$, and it maps each element $\Omega(d)$ of $\mathscr{B}$ to $\Gamma(d)$.

The remainder of this section is devoted to constructing a chain isomorphism

$$\Psi: (C, \Omega) \to (C, \Delta)$$

in the case when (C, Δ) is free. In the concluding remarks of Section IV.3 it was shown what modification is required in the definition of Ω for Ψ to be an isomorphism in the case when (C, Δ) is not free.

Just as in Chapters II and III, where Ψ was defined as a function which replaces the product $\times$ by a new product $\wedge$, so in this chapter Ψ will replace $\times$ by a new binary operation $\circ$, called the block product.

DEFINITION IV.3. Let k and l be blocks of dimension κ and λ, respectively, in an element of the language $\mathscr{C}$ of transpositions. (i.e. These blocks contain $\kappa - 1$ and $\lambda - 1$ appearances, respectively, of the symbol $\times$). Then, the block product $\kappa \circ l$ is defined by

$$k \circ l = k \times l - (-1)^{\kappa\lambda} l \times k.$$

Furthermore, we extend this product to linear combinations of blocks by linearity; i.e.

$$(j + k) \circ l = j \circ l + k \circ l$$

and

$$(-j) \circ k = -(j \circ k) = j \circ (-k).$$

The product in the above definition is non-associative. Consequently, an expression of the form,

$$e_1 \circ e_2 \circ \ldots \circ e_\kappa$$

which is obtained by replacing $\times$ by $\circ$ in a block of dimension κ, is meaningless, unless we adopt a *bracketing convention*. It will be useful to have such a convention in the special case when $e_1 \times e_2 \times \ldots \times e_\kappa$ is prime, and it will be based on the following theorem:

THEOREM IV.2. *Any prime block of dimension greater than* 1 *can be factored into two prime blocks.*

Proof. A prime block of dimension greater than 1 can be factored into the form $j \times j^*$ with j prime, because it is true if j is taken to be of dimension 1. If j^* is also prime, then the theorem is proved. Otherwise, let $k \times l$ be a factorization of j^*, such that k is the factor of lowest dimension such that

$$k > l \text{ or } k \simeq l.$$

This factorization exists, because j^* is not prime. If we can prove that $j \times k$ is prime, then it will have been shown that, whenever j is prime and j^* is not, j can be replaced by a larger prime factor and j^* by a smaller (non-void) factor, and enough repetitions of this process of enlarging j must eventually lead to a factorization which proves the theorem. Therefore, it remains to be shown that $j \times k$ is prime. We know that $j < k$, because $j < k \times l$ and either $k \times l < k$ or $k \times l \simeq k$. To prove that $j \times k$ is prime, consider every other factorization of it into two factors and show them to be in increasing order. There are two cases:
  (1) Let $k = k_1 \times k_2$.

$$j \times k_1 \times k_2 = j \times k < k < k_2.$$

Therefore,

$$j \times k_1 < k_2.$$

(ii) Let

$$j = j_1 \times j_2.$$

$$j_1 < j = j_1 \times j_2 < j_1 \times j_2 \times k.$$

Therefore,

$$j_1 < j_2 \times k.$$

This proves the theorem.

The above theorem provides a way of bracketing a κ-dimensional prime block h,

$$h = e_1 \times e_2 \times \ldots \times e_\kappa$$

so as to reduce it to a sequence of binary multiplications, such that each one of them is the product of two prime intervals within h, the first being less than the second in the natural lexicographic ordering. Let us define the following as the *conventional bracketing* of h: Separate h into two prime factors, and, if it can be done in more than one way, choose the one in which the left factor has the highest dimension. These two factors are set off by brackets, and the same process is applied again to the contents of each bracket. This is continued until h has been completely reduced to binary multiplications. Accordingly, the expression

$$e_1 \circ e_2 \circ \ldots \circ e_\kappa,$$

which involves a non-associative product, will be understood as having the *conventional bracketing*, as defined for h.

LEMMA IV.5. *If $e_1 \times e_2 \times \ldots \times e_\kappa$ is a κ-dimensional prime block, then it is the smallest block in the expansion of $e_1 \circ e_2 \circ \ldots \circ e_\kappa$ into a linear combination of blocks.*

Proof. Let $e_1 \times \ldots \times e_\kappa$ have the conventional bracketing. Starting with the outermost brackets and working in, replace the ×'s one-at-a-time by ○. At each stage we obtain an expression which has twice as many terms as the expression before. Going from one stage to the next, each term is replaced by a copy of itself of the form

$$\ldots \circ (j \times k) \circ \ldots$$

and a new term

$$\ldots \circ (k \times j) \circ \ldots ,$$

where $j < k$. If $e_1 \times \ldots \times e_\kappa$ is less or equal to every term in the first of these two expressions, then it is less than every term in the second. This is all we need for an inductive proof that $e_1 \times \ldots \times e_\kappa$ is less than or equal to every term in $e_1 \circ \ldots \circ e_\kappa$.

LEMMA IV.6. *If $e_1 \times e_2 \times \ldots \times e_\nu$ is a block of dimension ν, then*

$$\Delta(e_1 \circ e_2 \circ \ldots \circ e_\nu) = 0.$$

Proof. It suffices for a proof by induction to show that for

$$u = e_1 \circ e_2 \circ \ldots \circ e_\kappa$$

and

$$v = f_1 \circ f_2 \circ \ldots \circ f_\lambda,$$

if $\Delta(u) = 0$ and $\Delta(v) = 0$, then $\Delta(u \circ v) = 0$.

First consider blocks k and l which are terms in the expansions of u and v, respectively. Then

$$\Delta(k \times l) = (-1)^{\kappa}(k, l) + (-1)^{\kappa\lambda+\lambda}(l, k) + \ldots,$$

in which we have omitted writing all the terms which will cancel out in the expansion of $\Delta(u \times v)$. The verification of these cancellations is omitted, being a consequence of $\Delta(u) = 0$ and $\Delta(v) = 0$.

We are left with

$$\Delta(v \times u) = (-1)^{\kappa}(u, v) + (-1)^{\kappa\lambda+\lambda}(v, u)$$

and similarly

$$\Delta(v \times u) = (-1)^{\lambda}(v, u) + (-1)^{\kappa\lambda+\kappa}(u, v).$$

Therefore,

$$\Delta(u \circ v) = (-1)^{\kappa}(u, v) + (-1)^{\kappa\lambda+\lambda}(v, u) -$$
$$- (-1)^{\kappa\lambda+\lambda}(v, u) - (-1)^{2\kappa\lambda+\kappa}(u, v) = 0$$

This proves the lemma.

LEMMA IV.7. *For any block h in homological form there exists a linear combination h^0 of elements of $\mathscr{C}$ such that*
(i) $\Delta(h^0) = 0$.
(ii) *h is the smallest term in h^0 with a non-zero coefficient.*
(iii) *h is the only term in h^0 which is in homological form.*

Proof. If h is not of the form $s \times s$, and, if it is expressed by

$$h = e_1 \times e_2 \times \ldots \times e_\lambda,$$

where λ is its dimension, then let

$$h' = e_1 \circ e_2 \circ \ldots \circ e_\lambda.$$

If h is of the form $s \times s$, then let

$$h' = s' \times s'.$$

From Lemma IV.5, it is clear that h is the smallest term of h'. It is possible, however, that h' will contain terms other than h which are in homological form. If not,

$$h^0 = h',$$

and, if so, we take the smallest such term s and define h'' as

$$h'' = h' - s'.$$

This procedure is repeated until an expression is arrived at, which contains no terms in homological form other than h. This expression defines h^0 and will always be reached in a finite number of steps, because every term s, which is used, must be larger than the previous one. This proves conclusion (iii) of the lemma. The other conclusions follow from Lemmas IV.5 and IV.6 without difficulty.

LEMMA IV.8. *There is a chain isomorphism* Ψ *of the chain complex* (C, Ω) *onto the largest free subcomplex of* (C, Δ), *which is defined as follows on the standard basis* $(\mathcal{H}, \mathcal{D}, \mathcal{B})$ *of* (C, Ω):

(i) *For any element* $(h, i, \ldots)$ *of* $\mathcal{H}$, *expressed as a sequence of blocks,*
$$\Psi(h, i, \ldots) = (h^0, i^0, \ldots).$$

(ii) *For any element* $(h, i, \ldots, j \times k, l, \ldots)$ *of* $\mathcal{D}$, *in which* $(h, i, \ldots, j)$ *is the homology part,*

$$\Psi(h, i, \dots, j \times k, l, \dots) = (h^0, i^0, \dots, j^0 \times k, l, \dots).$$

(iii) *For any element* $\Omega(p)$ *of* $\mathscr{B}$

$$\Psi(\Omega(p)) = \Delta(\Psi(p)).$$

Proof. Ψ has been constructed such that, if it is a group isomorphism of C, then it is bound to be a chain isomorphism of (C, Ω), as well. Therefore, it is enough to show that Ψ is characterized by a $\gamma \times \gamma$ matrix with units on the main diagonal and zero below, where γ is the rank of C. This is accomplished by assigning a linear ordering to the basis $\mathscr{C}$, such that, if p is in $\mathscr{C}$, then $+ p$ or $- p$ is the smallest term with a non-zero coefficient in $\Psi(p)$. That is, when $\Psi(p)$ is expressed as a linear combination of elements of $\mathscr{C}$, p is smallest relative to the following ordering of $\mathscr{C}$.

Separate $\mathscr{C}$ into $\mathscr{B}$ and $\mathscr{D} \cup \mathscr{H}$, and define all elements of $\mathscr{B}$ as being less that all element of $\mathscr{D} \cup \mathscr{H}$. Next, place $\mathscr{B}$ in reverse lexicographic order and $\mathscr{D} \cup \mathscr{H}$ in lexicographic order. Here, the word *lexicographic* is used in the following sense:

The elements, being ordered, are sequences of blocks. Therefore, to order them lexicographically means to regard the set of all blocks as an alphabet, which in turn is placed in the natural lexicographic order. For the alphabet to be linearly ordered, it is necessary to assign a linear ordering to sets of the form

$$\{s, s \times s, s \times s \times s, \dots\},$$

containing equivalent blocks. Our proof is independent of how this is done.

The set $\mathscr{C}$ is such that no element, regarded as a sequence of blocks, is a partial sequence of another. Therefore, it is not necessary to specify whether the lexicographic ordering (relative to the alphabet of all blocks) is natural or common.

Suppose $p \in \mathscr{B}$, then let us establish that it is the smallest term of $\Psi(p)$ having a non-zero coefficient. Let p be expressed as a sequence of blocks

$$p = (h, i, \ldots, j, k, \ldots),$$

in which $(h, i, \ldots, j)$ is the homology part. Then,

$$\Psi(p) = \Delta\Psi\Omega^{-1}(p) = \Delta(h^0, i^0, \ldots, j^0 \times k, \ldots)$$

This contains terms in four categories:
  (i) $h^0, i^0, \ldots, j^0, k, \ldots$
  (ii) $h^0, i^0, \ldots, j^0 \times k_1, k_2, \ldots$
  (iii) $h^0, i^0, \ldots, k_1, j^0 \times k_2, \ldots$
  (iv) $h^0, i^0, \ldots, k, j^0, \ldots$
All the superscripts may be dropped except in that j^0 which appears in categories (iii) and (iv), because the omitted terms are all in $\mathscr{D}$. (p is less than all elements of $\mathscr{D}$). Then what remains in category (i) is p, and it is less than the expressions in the other categories, because the block j is larger than the blocks, $j \times k_1, k_1$, and k in the natural lexicographic ordering of blocks.

Suppose $p \in \mathscr{D} \cup \mathscr{H}$. Then $\Psi(p)$ contains no terms from $\mathscr{B}$, and p is the smallest term by virtue of Lemma IV.6.

Therefore, we have shown that Ψ can be expressed by a matrix with non-zero entries on the main diagonal and zeros below. When diagonal entries are not units, then (C, Ω) is isomorphic with a maximal free subcomplex of (C, Δ); otherwise (C, Δ) is free and (C, Ω) is isomorphic to it.

This proves the lemma, and also it proves the following theorem.

THEOREM IV.3. *A standard basis* $(\mathscr{H}', \mathscr{D}', \mathscr{B}')$ *for the largest free subcomplex of* (C, Δ) *is as follows:*

(i) $\mathcal{H}'$ is the set consisting of every element

$$(j_1^0, j_2^0, \ldots)$$

of C, where $(j_1, j_2, \ldots)$ is any element of $\mathscr{C}$ in homological form.

(ii) $\mathscr{D}'$ is the set consisting of every element

$$(j_1^0, j_2^0, \ldots, j_\theta^0 \times k_1, k_2, \ldots),$$

and

(iii) $\mathscr{B}'$ is the set consisting of every element

$$(j_1^0, j_2^0, \ldots, j_\theta^0, k_1, k_2, \ldots),$$

such that in both (ii) and (iii) $(j_1, j_2, \ldots, j_\theta)$ is the homological part of some element

$$(j_1, j_2, \ldots, j_\theta, k_1, k_2, \ldots) \in \mathscr{C}.$$

IV.6. A PROGRAM FOR SEQUENCES

The groups H and D are now completely determined for each of the complexes (C, Δ), (C, Ω), and (C, Γ). Therefore, we have determined the set of all programs in each complex which list a basis for C.

In particular, $\mathcal{H}$ and $\mathscr{D}$ are the input and instructions, respectively, of a graphical program in (C, Γ), which lists without repetitions the elements of $\mathscr{C}$. Every instruction which lists an element of $\mathscr{C}_1$, does so by transposing a pair of adjacent blocks in a previously determined element of $\mathscr{C}_1$. $\mathscr{D}_2$ is the set of all such transpositions, but in higher dimensions $\mathscr{D}$ does not generally consist of transpositions. For instance, under the interpretation Γ, the instruction $b \times c \times a$, belonging to $\mathscr{D}_3$, depends on $c \times a$, b and produces $b \times c, a$. That is,

$$\Gamma(b \times c \times a) = -c \times a, b + b \times c, a.$$

In order to determine a program explicitly, having $\mathscr{H}$ and $\mathscr{D}$ as its input and instructions, it is necessary to find a suitable linear ordering of the instructions, which will now be done by the inductive method described in Section I.3.

A subroutine $\mathscr{L}$ is used, which assigns to each element x of $\mathscr{C}$ the list $\mathscr{L}(x)$ of all instructions which depend on x. Then the program is specified by substituting this $\mathscr{L}$ into flowchart 2 of the appendix. It is easy to see from the definition of Γ in Section IV.4 how the subroutine $\mathscr{L}$ is to be constructed. For any x in $\mathscr{C}$ it is necessary for $\mathscr{L}$ to list all p in $\mathscr{C}$, such that x will be the negative term of $\Gamma(p)$. A subroutine to do this is given by part L of flowchart 8 in the appendix. The entire program in (C, Γ) for listing the elements of $\mathscr{C}$, is specified by flowcharts 2 and 8 of the appendix, taken together. The part of this program which merely lists the elements of $\mathscr{C}_1$ (all permutations of a given sequence) is specified by flowcharts 1 and 7 of the appendix, taken together.

IV.7. SAMPLE PROGRAMS

To demonstrate the kind of program, developed in this chapter, let us consider several examples. Each example calls for a full list without repetitions of some set.

EXAMPLE IV.1. Use a program in (C, Γ) to list all permutations of the sequence (a, b, b, c), each instruction in the program being a transposition of two adjacent terms of some sequence already on the list. The set $\mathscr{C}_1$, being listed, obviously has 12 elements. The minimum set $\mathscr{H}_1$ of input has one element, for which we use the sequence (a, b, b, c), whose terms are in alphabetical order. This choice is based on the construction of $\mathscr{H}$ in Section IV.3, and $\mathscr{H}_1$ is defined as the set of one-dimensional elements in $\mathscr{H}$.

Starting with this input, we use flowcharts 1 and 7 of the appendix to generate $\mathscr{C}_1$. Each line below represents one passage through loop in flowchart 1, and the column on the right is the desired list.

		$\mathscr{D}_2$	$\mathscr{C}_1$
$\mathscr{H}_1$	$=$		a, b, b, c
$\mathscr{L}(a, b, b, c)$	$=$	$\begin{cases} b \times a, b, c \to b, a, b, c \\ a, b, c \times b \to a, b, c, b \end{cases}$	
$\mathscr{L}(b, a, b, c)$	$=$	$b, b \times a, c \to b, b, a, c$	
$\mathscr{L}(a, b, c, b)$	$=$	$\begin{cases} b \times a, c, b \to b, a, c, b \\ a, c \times b, b \to a, c, b, b \end{cases}$	
$\mathscr{L}(b, b, a, c)$	$=$	$b, b, c \times a \to b, b, c, a$	
$\mathscr{L}(b, a, c, b)$	$=$	$b, c \times a, b \to b, c, a, b$	
$\mathscr{L}(a, c, b, b)$	$=$	$c \times a, b, b \to c, a, b, b$	
$\mathscr{L}(b, b, c, a)$	$=$	$b, c \times b, a \to b, c, b, a$	
$\mathscr{L}(b, c, a, b)$	$=$	$c \times b, a, b \to c, b, a, b$	
$\mathscr{L}(c, a, b, b)$	$=$	none	
$\mathscr{L}(b, c, b, a)$	$=$	$c \times b, b, a \to c, b, b, a$	
$\mathscr{L}(c, b, a, b)$	$=$	none	
$\mathscr{L}(c, b, b, a)$	$=$	none	

EXAMPLE IV.2. Use a program in (C, Γ) to list without repetitions all the elements of the language $\mathscr{C}$, associated with the sequence (a, b, c). $\mathscr{C}$ separates into 3 kinds of elements,

$$\mathscr{C} = \mathscr{C}_1 \cup \mathscr{C}_2 \cup \mathscr{C}_3,$$

where $\mathscr{C}_1$ is the set of permutations of (a, b, c), $\mathscr{C}_2$ is the set of transposition instructions available to generate $\mathscr{C}_1$, and $\mathscr{C}_3$ is the set of all instructions available to generate $\mathscr{C}_2$. Any program in the language $\mathscr{C}$ requires a minimal set $\mathscr{H}$ of input elements, where $\mathscr{H}$ has the separation

$$\mathcal{H} = \mathcal{H}_1 \cup \mathcal{H}_2 \cup \mathcal{H}_3,$$

induced by the separation of $\mathcal{C}$. In accordance with Section IV.3 the input consists of the following 6 elements:

$$\mathcal{H}_1 = \{(a, b, c)\}$$
$$\mathcal{H}_2 = \{(a, b \times c), (a \times b, c), (a \times c, b)\}$$
$$\mathcal{H}_3 = \{(a \times b \times c), (a \times c \times b)\}.$$

Starting with this input, we follow flowcharts 2 and 8 of the appendix. Each line in the format below represents one passage through the inner loop of the flowchart, but no line is written when $\lambda = 0$. The lines are indented in such a way that the elements of $\mathcal{C}_1$, $\mathcal{C}_2$, and $\mathcal{C}_3$ appear in separate columns, one for each passage through the outer loop. An arrow $\rightarrow$ runs from each instruction to the element it produces.

		$\mathcal{C}_3$	$\mathcal{C}_2$	$\mathcal{C}_1$
$\mathcal{H}_1$	=			a, b, c
$\mathcal{L}(a, b, c)$	=		$\begin{cases} b \times a, c \rightarrow b, a, c \\ a, c \times b \rightarrow a, c, b \end{cases}$	
$\mathcal{L}(b, a, c)$	=		$b, c \times a \rightarrow b, c, a$	
$\mathcal{L}(a, c, b)$	=		$c \times a, b \rightarrow c, a, b$	
$\mathcal{L}(b, c, a)$	=		$c \times b, a \rightarrow c, b, a$	
$\mathcal{H}_2$	=		$\begin{cases} a, b \times c \\ a \times b, c \\ a \times c, b \end{cases}$	
$\mathcal{L}(b \times a, c)$	=	$c \times b \times a \rightarrow$	$c, b \times a$	
$\mathcal{L}(c \times a, b)$	=	$b \times c \times a \rightarrow$	$b \times c, a$	
$\mathcal{L}(a \times b, c)$	=	$c \times a \times b \rightarrow$	$c, a \times b$	
$\mathcal{L}(a \times c, b)$	=	$b \times a \times c \rightarrow$	$b, a \times c$	
$\mathcal{H}_3$	=	$\begin{cases} a \times b \times c \\ a \times c \times b \end{cases}$		

EXAMPLE IV.3. Use a program in (C, Γ) to list all the elements of the language $\mathscr{C}$, associated with the sequence

$$(a, a, a, a, a, a).$$

This sequence is the only element of $\mathscr{C}_1$, and the elements of $\mathscr{C}_2$ are the 6 trivial transpositions:

$$(a \times a, a, a, a, a)$$

$$(a, a \times a, a, a, a)$$

$$. . .$$

$$(a, a, a, a, a \times a)$$

Nevertheless, the program for listing every element of $\mathscr{C}$ is not trivial. Let us abbreviate the expressions

$$a, a \times a, a \times a \times a, . . .$$

by the integers 1, 2, 3, ... Then, in accordance with Section IV.3, the input $\mathscr{H}$ for our program consists of 2 elements:

$$\mathscr{H} = \{(1, 1, 1, 1, 1, 1), (2, 1, 1, 1, 1, 1)\}$$

Starting with this input, we follow flowcharts 2 and 8 of the appendix, and obtain the following result:

	$\mathscr{C}_6$	$\mathscr{C}_5$	$\mathscr{C}_4$	$\mathscr{C}_3$	$\mathscr{C}_2$	$\mathscr{C}_1$
$\mathscr{H}_1 =$						(1, 1, 1, 1, 1, 1)
$\mathscr{H}_2 =$					(2, 1, 1, 1, 1)	
$\mathscr{L}(2, 1, 1, 1, 1) =$				(3, 1, 1, 1) →	(1, 2, 1, 1, 1, 1)	
$\mathscr{L}(1, 2, 1, 1, 1) =$				(1, 3, 1, 1) →	(1, 1, 2, 1, 1, 1)	
$\mathscr{L}(1, 1, 2, 1, 1) =$				(1, 1, 3, 1) →	(1, 1, 1, 2, 1, 1)	
$\mathscr{L}(1, 1, 1, 2, 1) =$				(1, 1, 1, 3) →	(1, 1, 1, 1, 1, 2)	
$\mathscr{L}(3, 1, 1, 1) =$			(4, 1, 1) →	(2, 2, 1, 1)		
$\mathscr{L}(2, 2, 1, 1) =$			(2, 3, 1) →	(2, 1, 2, 1)		
$\mathscr{L}(2, 1, 2, 1) =$			{ (3, 2, 1) → (2, 1, 3) →	(1, 2, 2, 1) (2, 1, 1, 2)		
$\mathscr{L}(2, 1, 1, 2) =$			(3, 1, 2) →	(1, 2, 1, 2)		
$\mathscr{L}(1, 2, 1, 2) =$			(1, 3, 2) →	(1, 1, 2, 2)		
$\mathscr{L}(4, 1, 1) =$		(5, 1) →	(1, 4, 1)			
$\mathscr{L}(2, 1, 3) =$		(3, 3) →	(1, 2, 3)			
$\mathscr{L}(3, 1, 2) =$		(4, 2) →	(2, 2, 2)			
$\mathscr{L}(1, 4, 1) =$		(1, 5) →	(1, 1, 4)			
$\mathscr{L}(5, 1) =$	(6) →	(2, 4)				

Here, $\mathscr{C}$ is simply the set of ordered partitions of 6, and the program lists them by numbers of parts.

Chapter V

Bracketings

V.1. THE LANGUAGE

This chapter is concerned with how to make a list of all the ways to bracket the terms of a finite sequence. The *bracketing convention* is that each pair of brackets encloses a proper subinterval of the sequence, distinct from the subinterval enclosed by any other pair of brackets, and any two intervals, so enclosed, are either separate or nested. A proper subinterval consists of a first term and a last term, which are distinct from each other, and of every term in between, but not of the whole sequence. Every sequence has an *empty bracketing* which means that none of its proper subintervals has been enclosed by brackets, and every sequence with three or more terms has other bracketings as well.

To begin with, let us consider how to list without repetitions every element of the set $\mathscr{C}_1$ of complete bracketings of a given sequence s. A bracketing is *complete* when no further bracketing is possible, which means that, if s has μ terms, then there are $(\mu - 2)$ pairs of brackets in it. The list will be made by means of a program, whose instructions are taken from a set $\mathscr{C}_2$ of rebracketings, defined as follows: To *rebracket* an element of $\mathscr{C}_1$ means to remove a pair of brackets and replace it by the only other possible pair of brackets. Each element p of $\mathscr{C}_2$ is denoted by an element of $\mathscr{C}_1$ with one pair of brackets removed. The element p is capable of exactly two non-empty bracketings, q and r, each being an element of $\mathscr{C}_1$. If the pair of bracketings in q and the pair in r, which differ, are such that those in q stand to the left of those in r, then let us adopt the convention that $+p$ is the instruction which converts q into r and $-p$ is the instruction which converts r into q.

As an example, let us consider the case in which s has 4 terms. Since, from the standpoint of bracketing, it makes no difference what the terms are, let us express it as follows:

$$s = aaaa.$$

Then $aa(aa)$ is an example of an instruction in $\mathscr{C}_2$. It tells us to make the following conversion of one element of $\mathscr{C}_1$ to another:

$$(aa)(aa) \rightarrow a(a(aa)).$$

In other words, it tells us to shift one pair of brackets to the right.

For any s it will be shown that there exist programs which list every element of $\mathscr{C}_1$, using instructions of the above type and using only one input element from $\mathscr{C}_1$. Each instruction depends on the input or on an element produced by a previous instruction. Furthermore, if $\mathscr{C}_{\pi+1}$ is defined as the set of bracketings of s which are formed by removing π pairs of brackets from elements of $\mathscr{C}_1$, then there exist programs in $\mathscr{C}_{\pi+1}$ which list the elements of $\mathscr{C}_\pi$. For $\pi > 1$ the program, which lists the elements of $\mathscr{C}_\pi$, will use no input other than the instructions required to list all the elements of $\mathscr{C}_{\pi-1}$. Therefore, it follows by induction that there exist programs which list the set $\mathscr{C}$ of all bracketings of s, using only one input element from $\mathscr{C}_1$ and using instructions from $\mathscr{C}$ itself. Our object is to define the set of all such programs and to construct a sample one. The theory, developed in Chapter I, lends itself to this objective.

The *language of bracketing* for a μ-termed sequence will be expressed in the form

$$\mathscr{C} = \mathscr{C}_1 \cup \mathscr{C}_2 \cup \ldots \cup \mathscr{C}_\pi \cup \ldots \cup \mathscr{C}_{\mu-1}$$

where

(i) $\mathscr{C}_{\mu-1}$ is set containing the single unbracketed expression $aa \ldots a$, in which a appears μ times.

(ii) $\mathscr{C}_\pi$ is the set of all elements formed by placing one pair of brackets in an element of $\mathscr{C}_{\pi+1}$ in accordance with the bracketing convention, defined above.

The induction, by which $\mathscr{C}$ has been defined, cannot go beyond the completely bracketed expressions in $\mathscr{C}_1$. In the next three sections $\mathscr{C}$ is made into a language in three different, but isomorphic ways, by assigning interpretations to its elements as instructions. For example, the interpretation, mentioned above, of $aa(aa)$ as an instruction would be expressed by the equation,

$$\Gamma(aa(aa)) = -(aa)(aa) + a(a(aa)).$$

The negative term on the right is what the instruction *depends on* and the positive term is what it *produces*.

It is a simple exercise in combinatorics to determine that the number of elements in the set $\mathscr{C}_\pi$ is

$$\binom{2\mu - \pi - 1}{\mu}\binom{\mu - 2}{\pi - 1}\frac{1}{\mu - \pi},$$

where μ is the number of terms in the sequence being bracketed. Our object is to list without repetitions the elements of this set for each π.

V.2. THE INTERPRETATION Δ

The interpretation Δ of $\mathscr{C}$ is defined by embedding $\mathscr{C}$ in a chain complex (C, Δ), in which C is the free Abelian group generated by

$\mathscr{C}$, and Δ is the boundary function, which is determined by giving a value to $\Delta(p)$ for each $p \in \mathscr{C}$, as follows:

(i) If $p \in \mathscr{C}_1$, then $\Delta(p) = 0$.

(ii) If $p \in \mathscr{C}_\pi$, then $\Delta(p) = \Sigma (- 1)^\theta q$,

where the sum is taken over every $q \in \mathscr{C}_{\pi - 1}$ which is obtainable by bracketing p, and θ is the total number of a's and left brackets in q up to and including (as we read q from left to right) the left bracket which was put into p to form q.

The number θ in the above interpretation of $\mathscr{C}$ has been chosen such that

$$\Delta\Delta = 0.$$

It is clear that, in expanding $\Delta\Delta(p)$ into a linear combination of elements of $\mathscr{C}_{\pi - 2}$, each term q is formed by putting two pairs of brackets into p. For each term q, so obtained, there will be a term $- q$, obtained by inserting the same brackets in the opposite order. This implies that for all p

$$\Delta\Delta(p) = 0,$$

and that (C, Δ) is a chain complex.

V.3. THE INTERPRETATION Ω

A chain complex (C, Ω), having a standard basis $(\mathscr{H}, \mathscr{D}, \mathscr{B})$, is to be constructed, which will be shown to be isomorphic to (C, Δ).

Let p be any element in the language $\mathscr{C}$ of bracketing. We regard p as being either equal to the sequence $aaa \ldots a$ or obtainable from it by successive bracketing. Let us define a pair of brackets in p as an *inner pair*, if it does not embrace first a in the sequence, and let us define an inner pair as being the first *inner pair*, if its left bracket stands to the left of every inner pair which is either contained in p or could be put into p.

(i) $p \in \mathscr{H}$, if p does not contain an inner pair of brackets and cannot have an inner pair put into it.

It is evident that there is only one bracketing of $aaa \ldots a$ which belongs to $\mathscr{H}$. It is

$$(\ldots(((aa)a)a)\ldots a)a,$$

in which every pair of brackets embraces the first a and no further brackets can be put in.

(ii) $p \in \mathscr{D}$, if p does not contain a first inner pair of brackets, but could have such a pair put in.

(iii) $p \in \mathscr{B}$, if p contains a first inner pair of brackets.

It is evident that there is at most one way to put a first inner pair of brackets into any p. Consequently, if p is in $\mathscr{D}$, we can put the first inner pair of brackets into it and denote the newly bracketed expression by $\Omega(p)$. Thus, we define a one-to-one function Ω of $\mathscr{D}$ onto $\mathscr{B}$. If this function is given the value zero on all elements of $\mathscr{B} \cup \mathscr{H}$ and is extended to a linear transformation of C into itself, then it determines a chain complex (C, Ω) which has $(\mathscr{H}, \mathscr{D}, \mathscr{B})$ as a standard basis.

V.4. THE INTERPRETATION Γ

A chain complex (C, Γ) is now constructed, which is isomorphic to both (C, Δ) and (C, Ω). The main fact about this complex is that it contains a graphical program which lists without repetitions all the elements of $\mathscr{C}$. $\mathscr{H}$ and $\mathscr{D}$, as defined in the last section, are the input and instructions of such a program.

$\mathscr{H}$ and $\mathscr{D}$ are defined in Section V.3. For any p in $\mathscr{D}$ let ij be the part of it which is embraced by the first inner pair of brackets in $\Omega(p)$.

$$p = \ldots hij \ldots$$
$$\Omega(p) = \ldots h(ij) \ldots.$$

Any expression which is enclosed by brackets has two or more factors, and we define i as the first factor embraced by the first inner pair of brackets.

In $\Omega(p)$ there is a factor h which precedes the *first inner pair* of brackets. From the definition of first inner pair it follows that h is neither void, nor is it preceded by anything except possibly some left brackets. Therefore, the following equation defines Γ uniquely on the set $\mathscr{D}$:

$$\Gamma(\ldots hij \ldots) = -[\ldots (hi)j \ldots] + [\ldots h(ij) \ldots]$$

If we can verify that the set

$$\mathscr{H} \cup \mathscr{D} \cup \Gamma \mathscr{D}$$

constitutes a basis for C, then Γ will be completely defined by giving it the value zero on all elements of $\mathscr{H} \cup \Gamma \mathscr{D}$. To make this verification, all we need to do is to order the elements of $\mathscr{C}$, where

$$\mathscr{C} = \mathscr{H} \cup \mathscr{D} \cup \Omega \mathscr{D}$$

and those of

$$\mathscr{H} \cup \mathscr{D} \cup \Gamma \mathscr{D}$$

correspondingly, so that the square matrix which relates them has ones on the diagonal and zeros below. This is achieved by any ordering in which the elements of $\Omega \mathscr{D}$ precede those of $\mathscr{H} \cup \mathscr{D}$ and in which the elements of $\Omega \mathscr{D}$ have an ordering which increases with the number of brackets to the left of h. This ordering was chosen, so that the negative term of $\Gamma(p)$ would be larger than the positive term for all p in $\mathscr{D}$.

V.5. LANGUAGE ISOMORPHISMS

There is an obvious chain isomorphism from (C, Ω) to (C, Γ). It is the identity map, when it is restricted to $\mathscr{H} \cup \mathscr{D}$, and it takes each element $\Omega(d)$ of $\mathscr{B}$ to $\Gamma(d)$. Let us prove that the same is true when Γ is replaced by Δ.

LEMMA V.1. *A chain isomorphism of (C, Ω) to (C, Δ) is uniquely determined by the map which is the identity on $\mathscr{D} \cup \mathscr{H}$ and which on $\Omega\mathscr{D}$ carries each $\Omega(d)$ to $\Delta(d)$.*

Proof. All we need to do is to order the elements of $\mathscr{C}$, where

$$\mathscr{C} = \mathscr{H} \cup \mathscr{D} \cup \Omega\mathscr{D}$$

and those of

$$\mathscr{H} \cup \mathscr{D} \cup \Delta\mathscr{D}$$

correspondingly, so that the square matrix which connects them has units on the diagonal and zeros below. This is achieved by any ordering in which the elements of $\Omega\mathscr{D}$ precede those of $\mathscr{H} \cup \mathscr{D}$, and in which the elements of $\Omega\mathscr{D}$ have an ordering which increases with the total number of a's and left brackets which are to the left of the first inner pair of brackets of each element of $\Omega\mathscr{D}$. This ordering gives the desired result, because for any d in $\mathscr{D}$, when $\Delta(d)$ is expanded to a linear combination of elements of $\mathscr{C}$, the term $\Omega(d)$ appears with a unit coefficient and it precedes all the other terms in the linear order.

This proves the lemma and also the next theorem.

THEOREM V.1. *A standard basis $(\mathscr{H}, \mathscr{D}, \Delta\mathscr{D})$, for the complex*

(*C*, Δ) *of bracketings of aaa . . . a is as follows: Let* inner *brackets be ones which do not enclose the first a; then*

(i) $\mathscr{H}$ *contains only one element*

$$(. . . (((a)a)a)a . . .)a,$$

in which every non-inner bracketing appears.

(ii) $\mathscr{D}$ *consists of every bracketing p which contains a position in which a pair of inner brackets can be placed, such that the left bracket stands to the left of all inner brackets which are either contained in p or could be put into it.*

(iii) Δ$\mathscr{D}$ *is the set* {Δ(*d*): *d* ∈ $\mathscr{D}$} *of linear combinations of bracketings.*

V.6. A PROGRAM FOR BRACKETINGS

The groups *H* and *D* are now explicitly determined for each of the complexes (*C*, Δ), (*C*, Ω), and (*C*, Γ). Therefore, we have determined the set of all programs in each complex, which list a basis for *C*.

In particular, $\mathscr{H}$ and $\mathscr{D}$ are the input and instructions of a graphical program in (*C*, Γ), which lists without repetitions, the elements of $\mathscr{C}$. The instructions have to be ordered in such a way that the element, which each instruction depends on, is in $\mathscr{H}$ or was produced by a previous instruction. This will now be done by the inductive method described in Section I.3.

A subroutine $\mathscr{L}$ is used, which assigns to each element *x* of $\mathscr{C}$, the list $\mathscr{L}(x)$ of all instructions which depend on *x*. Then the program is specified by substituting this $\mathscr{L}$ into flowchart 2 of the appendix. A flowchart for $\mathscr{L}$ is given by part *L* of flowchart 10 in the appendix, and an entire program in (*C*, Γ) for listing the elements of $\mathscr{C}$ is specified by flowcharts 2 and 10 of the appendix, taken together. The part of this program which merely lists the elements of $\mathscr{C}_1$ (the complete bracketings of $aaa \ldots a$) is specified by flowcharts 1 and 9 of the appendix, taken together.

V.7. SAMPLE PROGRAMS

EXAMPLE V.1. Use a program (C, Γ) to list without repetitions all the complete bracketings of $aaaaa$, each instruction being the shift of a single pair of brackets to the right. The single input element is $(((aa)a)a)a$, whose brackets are all the way to the left. Starting with this input, we use flowcharts 1 and 9 of the appendix to generate $\mathscr{C}_1$. Each $\mathscr{L}$ below represents one passage through the loop in flowchart 1. The instructions are in the middle column. To the left of each one is what it depends on (always taken from a preceding line), and to the right is what it produces.

	$\mathscr{D}_2$	$\mathscr{C}_1$
$\mathscr{H}_1 =$		$(((aa)a)a)a$
$\mathscr{L}[(((aa)a)a)a] =$	$((aa)a)aa \rightarrow ((aa)a)(aa)$	
	$((aa)aa)a \rightarrow ((aa)(aa))a$	
	$((aaa)a)a \rightarrow ((a(aa))a)a$	
$\mathscr{L}[((aa)a)(aa)] =$	$(aa)a(aa) \rightarrow (aa)(a(aa))$	
	$(aaa)(aa) \rightarrow (a(aa))(aa)$	
$\mathscr{L}[((aa)(aa))a] =$	$(aa)(aa)a \rightarrow (aa)((aa)a)$	
	$(aa(aa))a \rightarrow (a(a(aa)))a$	
$\mathscr{L}[((a(aa))a)a] =$	$(a(aa)a)a \rightarrow (a((aa)a))a$	
$\mathscr{L}[(aa)(a(aa))] =$	$aa(a(aa)) \rightarrow a(a(a(aa)))$	
$\mathscr{L}[(a(aa))(aa)] =$	$a(aa)(aa) \rightarrow a((aa)(aa))$	
$\mathscr{L}[(aa)((aa)a)] =$	$aa((aa)a) \rightarrow a(a((aa)a))$	
$\mathscr{L}[(a(a(aa)))a] =$	$a(a(aa))a \rightarrow a((a(aa))a)$	
$\mathscr{L}[(a((aa)a))a] =$	$a((aa)a)a \rightarrow a(((aa)a)a)$	

Notice that only one instruction depends on $((a(aa))a)a$, even though it has two non-inner pairs of brackets which could be removed, because the bracketing obtained by removing the outermost pair of brackets is not an element of $\mathcal{D}_2$.

EXAMPLE V.2. This is a continuation of the last example. Still using a program in (C, Γ), let us list the bracketings of $aaaaa$, not already listed. No further input is needed. We follow flowcharts 2 and 10 of the appendix. Since it begins by repeating all the steps of Example V.1, this part is not repeated below.

	$\mathcal{D}_3$	$\mathcal{B}_2$
$\mathcal{L}[((aa)a)aa] =$	$\begin{cases} (aa)aaa \rightarrow (aa)(aaa) \\ (aaa)aa \rightarrow (a(aa))aa \end{cases}$	
$\mathcal{L}[((aa)aa)a] =$	$(aaaa)a \rightarrow (a(aaa))a$	
$\mathcal{L}[(aa)a(aa)] =$	$aaa(aa) \rightarrow a(aa(aa))$	
$\mathcal{L}[(aa)(aa)a] =$	$aa(aa)a \rightarrow a(a(aa)a)$	
$\mathcal{L}[(aa)(aaa)] =$	$aa(aaa) \rightarrow a(a(aaa))$	
$\mathcal{L}[(a(aa))aa] =$	$a(aa)aa \rightarrow a((aa)aa)$	
$\mathcal{L}[(a(aaa))a] =$	$a(aaa)a \rightarrow a((aaa)a)$	
	$\mathcal{C}_4$	$\mathcal{B}_3$
$\mathcal{L}[(aa)aaa] =$	$aaaaa \rightarrow a(aaaa)$	

This completes the list of all bracketings of $aaaaa$, started in Example V.1.

Chapter VI

Clustering

VI.1. THE LANGUAGE

This chapter is concerned with how to make a list of all the ways of clustering the elements of a finite set, which means recognizing a family of non-void subsets, called clusters, any pair of which is either disjoint or nested. The list of clusterings is greatly shortened, without loss of generality, if we define a *cluster* as a subset with more than one element and not equal to the whole set. This reduces the number of clusterings of a set of μ elements, by a factor of $2^{\mu+1}$, since we have eliminated $\mu + 1$ subsets, each of which could otherwise have been either a cluster or not. Accordingly, a *clustering* is defined as a family of such clusters, any two of them being disjoint or nested. This includes the case of a clustering whose family of clusters is void.

This chapter is an extension of Chapter V, by virtue of the fact that a clustering of a set can be expressed by a bracketed sequence, where each cluster consists of the terms between a pair of brackets. For example, the clusterings of the set $\{a, b, c\}$ can be expressed by $a(bc)$, $(ab)c$, $(ac)b$, and abc. Furthermore, this notation can be made to be unique, by giving the underlying set an alphabetical ordering and extending this ordering lexicographically to bracketed expressions. The bracketings of a μ-termed sequence correspond exactly to those clusterings of a set of μ elements whose terms appear in alphabetical order, when written in the unique manner, described above. In the above example the bracketings of *abc* are all clusterings of {*a, b, c*} except for (*ac*)*b*, whose letters are out of order. Therefore, regarding Chapter V as an introduction to this one, we may proceed directly to the embedding of the clusterings of a set in a suitable chain complex.

The *language of clustering* of the set

$$\{a_1, a_2, \ldots, a_\mu\}$$

is of the form

$$\mathscr{C} = \mathscr{C}_1 \cup \mathscr{C}_2 \cup \ldots \cup \mathscr{C}_\pi \cup \ldots \cup \mathscr{C}_{\mu-1}$$

where $\mathscr{C}_1$ is the set of complete clusterings each containing $\mu - 2$ clusters, and in general $\mathscr{C}_\pi$ is the set of clusterings each of which contains $\mu - \pi - 1$ clusters.

The number of elements in the set $\mathscr{C}_1$ of complete clusterings of μ elements is as follows:

$$\gamma(\mu, 1) = 1 \cdot 3 \cdot 5 \ldots (2\mu - 3).$$

This comes from the recursive formula

$$\gamma(\mu, 1) = (2\mu - 3)\gamma(\mu - 1, 1),$$

which follows from the fact that there of $2\mu - 3$ complete clusterings which can be formed when a_μ is adjoined to a complete clustering of the set $\{a_1, \ldots, a_{\mu-1}\}$.

Similarly, the number of elements in $\mathscr{C}_\pi$ can be found by the following recursive formula:

$$\gamma(\mu, \pi) = (2\mu - 2 - \pi)\gamma(\mu - 1, \pi) +$$
$$+ (\mu - \pi)\gamma(\mu - 1, \pi - 1)$$

Our object, however, is not to count the elements in $\mathscr{C}_\pi$, but to list them.

VI.2. THE INTERPRETATION Δ

The interpretation Δ of $\mathscr{C}$ is defined by embedding $\mathscr{C}$ in a chain complex (C, Δ), in which C is the free Abelian group generated by $\mathscr{C}$, and Δ is the boundary function, which is determined by giving a value $\Delta(p)$ for each $p \in \mathscr{C}$, as follows:

(i) If $p \in \mathscr{C}_1$, then $\Delta(p) = 0$.

(ii) If $p \in \mathscr{C}_\pi$, then $\Delta(p) = \Sigma \, (-1)^\theta q$, where the sum is taken over every $q \in \mathscr{C}_{\pi-1}$, which is obtainable by introducing one more cluster to p, and, θ is the number of left brackets in q, written as a bracketed sequence, up to and including (as we read from left to right) the left bracket of the cluster which was put into p to form q.

The above construction of Δ assumes that a clustering is written in the form of a bracketed expression, and the following rule is given, so that the value of θ may be independent of the manner in which the clustering is expressed by brackets: If two adjacent factors p and r, are re-ordered and they contain π and ρ left brackets, respectively, then

$$(\ldots pr \ldots) = (-1)^{\pi\rho}(\ldots rp \ldots).$$

When applying Δ to a bracketed sequence p, there will, in general, be summands of $\Delta(p)$ whose terms cannot be written in the same order as those of p. Hence, the above reordering rule is an essential part of the definition of Δ. For example

$$\Delta(a(bc)d) = -(a(bc))d - (ad)(bc) - a((bc)d)$$

The second term is obtained by first writing $a(bc)d$ as

$ad(bc)$ or $(bc)ad$,

then by putting brackets around ad and calculating the coefficient. This gives

$$- (ad)(bc) \text{ or } + (bc)(ad),$$

which are equal under the reordering rule. It is easy to verify, in general, that we get the same value when applying Δ to either side of the identity

$$(\ldots pr \ldots) = (-1)^{\pi\rho}(\ldots rp \ldots).$$

It is also easy to see that

$$\Delta\Delta = 0.$$

It follows that, if C is the free Abelian group generated by $\mathscr{C}$, then (C, Δ) constitutes a chain complex.

VI.3. THE INTERPRETATION Ω

A chain complex (C, Ω), having a standard basis $(\mathscr{H}, \mathscr{D}, \mathscr{B})$, is to be constructed, which will be shown to be isomorphic to (C, Δ).

Let p be any element of the language $\mathscr{C}$ of clusterings. Of course, $\mathscr{C}$ is merely the set of clusterings of $\{a_1, \ldots, a_\mu\}$, but we call it a 'language' since it is to be given an interpretation. Let us define a cluster in p as an *inner cluster*, if it does not include a_1 and let us define an inner cluster as being the *first inner cluster*, if, when written lexicographically as a bracketed sequence, it either includes or stands to the left of every inner cluster which is in p or could be included in it.

To say that a clustering p is expressed as a *lexicographically bracketed sequence* means the following: If p contains no clusters it is written as

$$a_1 a_2 \ldots a_\mu.$$

If a single cluster is introduced, its elements form a subsequence,

$$a_\phi \ldots,$$

which is removed from $a_1 a_2 \ldots a_\mu$ and placed where a_ϕ had been with brackets around it. This operation can be repeated, if there are more clusters. The obvious way is to put in the maximal clusters first. Then the elements of any subsequent cluster, to be introduced, form a subsequence

$$a_\psi \ldots$$

of an already bracketed interval, such as $(a_\phi \ldots)$. This subsequence is removed from $(a_\phi \ldots)$, placed in brackets, and returned to the position originally occupied by a_ψ. When every cluster has been repositioned and bracketed in this manner, the whole clustering will have been reduced to a lexicographically bracketed sequence.

(i) $p \in \mathscr{H}$, if p does not contain an inner cluster and could not have one put into it.

It is evident that there are $(\mu - 1)!$ clusterings in $\mathscr{H}$, each one being expressible as

$$(\ldots (((a_1 b_2) b_3) b_4) \ldots b_{\mu-1}) b_\mu,$$

where $b_2 b_3 \ldots b_\mu$ is a permutation of $a_2 a_3 \ldots a_\mu$.

(ii) $p \in \mathscr{D}$, if p does not contain a first inner cluster, but could have one put into it.

(iii) $p \in \mathscr{B}$, if p contains a first inner cluster.

It is evident that there is at most one way to put a first inner cluster into any p. Consequently, if p is in $\mathscr{D}$, we can put the first inner cluster into it and denote the new clustering, so obtained, by $\Omega(p)$. Thus we define a one-to-one function Ω of $\mathscr{D}$ onto $\mathscr{B}$. If this function is given the value zero on all elements of $\mathscr{B} \cup \mathscr{H}$ and is extended to a linear transformation of C into itself, then it determines a chain complex (C, Ω), which has $(\mathscr{H}, \mathscr{D}, \mathscr{B})$ as a standard basis.

VI.4. THE INTERPRETATION Γ

A chain complex (C, Γ) is now constructed, which is to be isomorphic to both (C, Δ) and (C, Ω), the main fact about this complex being that it contains a graphical program which lists without repetitions all the elements of $\mathscr{C}$. $\mathscr{H}$ and $\mathscr{D}$, as defined in the last section, are the input and instructions of such a program.

Let us define a *factor* of a clustering in $\mathscr{C}$ to be one of its maximal clusters or to be a singleton a_ϕ, provided that a_ϕ is not contained in any cluster. Each factor, which is not a singleton, is itself a clustering of some subset of $\{a_1, \ldots a_\mu\}$ and can in turn be factored. When a clustering is expressed as a lexicographically bracketed sequence, its factors will be a partition of it into intervals. For example, there are two factors in

$$(a(be)d)(cf)$$

the first of which has three factors and the second of which has two.

For any p in $\mathscr{D}$ let $\Gamma(p)$ be constructed as follows: By definition p is a clustering which does not contain a first inner cluster, but such a cluster can be put into it, which is how $\Omega(p)$ is formed. If $\Omega(p)$ is written as a lexicographically bracketed sequence, then it is of the form

$$\Omega(p) = \ldots h(f_1 f_2 f_3 \ldots) \ldots,$$

where $(f_1 f_2 f_3 \ldots)$ is the first inner cluster, reduced to its factors, and h is the factor which precedes it. (h is either a_1 or the largest cluster which contains a_1, and the only thing which can precede h in the above notation is left brackets.) Therefore,

$$p = \ldots h f_1 f_2 f_3 \ldots,$$

and we define Γ as follows:

$$\Gamma(p) = [\ldots h(f_1 f_2 f_2 \ldots) \ldots] - [\ldots (hf_1) f_2 f_3 \ldots].$$

As an example of this let

$$p = (a(be)d)(cf).$$

Then

$$\Omega(p) = (a((be)d))(cf).$$

Here, the first inner cluster is $((be)d)$, which has two factors, and

$$\Gamma(p) = (a((be)d))(cf) - ((a(be))d)(cf).$$

To complete the definition of Γ, we give it the value zero on all elements of

$$\mathscr{H} \cup \Gamma \mathscr{D}$$

and verify that the set

$$\mathscr{H} \cup \mathscr{D} \cup \Gamma \mathscr{D}$$

on which Γ is now defined, constitutes a basis for C. This verification may be made exactly as it was in Section V.4.

VI.5. LANGUAGE ISOMORPHISMS

There is an obvious chain isomorphism form (C, Ω) to (C, Γ), which is the identity map, when it is restricted to $\mathcal{H} \cup \mathcal{D}$, and which takes each $\Omega(d)$ to $\Gamma(d)$. Let us prove that the same is true for Δ as well as Γ.

LEMMA VI.1. *A chain isomorphism (C, Ω) to (C, Δ) is uniquely determined by the map which is the identity on $\mathcal{D} \cup \mathcal{H}$ and which on $\Omega \mathcal{D}$ carries each $\Omega(d)$ to $\Delta(d)$.*

Proof. All we need to do is to order the elements of $\mathcal{C}$, where

$$\mathcal{C} = \mathcal{H} \cup \mathcal{D} \cup \Omega \mathcal{D}$$

and those of

$$\mathcal{H} \cup \mathcal{D} \cup \Delta \mathcal{D},$$

correspondingly, so that the square matrix which connects them has units on the diagonal and zeros below. It is achieved by any ordering in which the elements of $\Omega \mathcal{D}$ precede those of $\mathcal{H} \cup \mathcal{D}$, and in which the elements of $\Omega \mathcal{D}$ have an ordering which increases with the number of non-inner clusters in each one. (A non-inner cluster is one which contains a_1.)

This ordering suffices, because of the fact that every term of $\Delta(p)$, which belongs to $\Omega \mathcal{D}$, has one more non-inner cluster than $\Omega(p)$. This property puts the matrix being considered, in the desired form.

This proves the lemma and also the next theorem.

THEOREM VI.1. *A standard basis $(\mathcal{H}, \mathcal{D}, \Delta \mathcal{D})$ for the complex (C, Δ) of clusterings of the set*

$$\{a_1, a_2, \ldots, a_\mu\}$$

is as follows:

(i) $\mathscr{H}$ *consists of the* $(\mu - 1)!$ *complete clusterings, each of the form*

$$(\ldots(((a_1 b_2) b_3) b_4)\ldots) b_\mu,$$

where $b_2 b_3 \ldots b_\mu$ *is a permutation of* $a_2 a_3 \ldots a_\mu$.

(ii) $\mathscr{D}$ *consists of every clustering which is formed by removing the first inner cluster from another clustering.*

(iii) $\Delta\mathscr{D}$ *is the set* $\{\Delta(d) : d \in \mathscr{D}\}$ *of linear combinations of clusterings.*

VI.6. A PROGRAM FOR CLUSTERINGS

The groups H and D are now explicitly determined for each of the complexes (C, Δ), (C, Ω), and (C, Γ). Therefore, we have determined the set of all programs in each complex, which list a basis for C.

In particular, $\mathscr{H}$ and $\mathscr{D}$ are the input and instructions of a graphical program in (C, Γ) which lists without repetitions all the elements of $\mathscr{C}$. The instructions have to be ordered in such a way that the element which each instruction depends on is in $\mathscr{H}$ or was produced by a previous instruction. This will now be done by the inductive method, described in Section I.3.

A subroutine $\mathscr{L}$ is used, which assigns to each element x of $\mathscr{C}$ the list $\mathscr{L}(x)$ of all instructions which depend on x. Then a program is specified by substituting this $\mathscr{L}$ into flowchart 2 of the appendix. A flowchart for $\mathscr{L}$ is given by part L of flowchart 12 in the appendix, and an entire program in (C, Γ) for listing the elements of $\mathscr{C}$ is specified by flowcharts 2 and 12 of the appendix, taken together. The part of this program which merely lists the elements of $\mathscr{C}_1$ (the complete clusterings of $\{a_1, \ldots, a_\mu\}$) is specified by flowcharts 1 and 11 of the appendix, taken together.

VI.7. SAMPLE PROGRAM

Using the recursive formula, given at the end of Section VI.1, we find that the cardinalities of the sets $\mathscr{C}_1, \mathscr{C}_2, \ldots, \mathscr{C}_{\mu-1}$ for small values of μ are as follows:

	$\mathscr{C}_1$	$\mathscr{C}_2$	$\mathscr{C}_3$	$\mathscr{C}_4$	$\mathscr{C}_5$
$\mu = 2$	1				
$\mu = 3$	3	1			
$\mu = 4$	15	10	1		
$\mu = 5$	105	105	25	1	
$\mu = 6$	945	1260	490	56	1

In view of the rapid increase of these numbers, let us limit ourselves to the single example, in which $\mu = 4$.

EXAMPLE VI.1. Use a program in (C, Γ) to list all the clusterings of the four letters a, b, c and d. The input consists of the clusterings formed by permuting the letters b, c, and d in the clustering $((ab)c)d$. (Notice that if we were to use only the single input element $((ab)c)d$ and ignore the other five, our program would reduce to the one discussed in Chapter V.) We follow flowcharts 2 and 12 of the appendix. (We would use flowcharts 1 and 11 to list $\mathscr{C}_1$ by itself.)

	$\mathscr{C}_3$	$\mathscr{C}_2$	$\mathscr{C}_1$
$\mathscr{H}_1 =$			$\begin{cases} ((ab)c)d \\ ((ab)d)c \\ ((ac)b)d \\ ((ac)d)b \\ ((ad)b)c \\ ((ad)c)b \end{cases}$
$\mathscr{L}(((ab)c)d) =$			$\begin{cases} (abc)d \rightarrow & (a(bc))d \\ (ab)cd \rightarrow & (ab)(cd) \end{cases}$
$\mathscr{L}(((ab)d)c) =$			$(abd)c \rightarrow (a(bd))c$
$\mathscr{L}(((ac)b)d) =$			$(ac)bd \rightarrow (ac)(bd)$
$\mathscr{L}(((ac)d)b) =$			$(acd)b \rightarrow (a(cd))b$
$\mathscr{L}(((ad)b)c) =$			$(ad)bc \rightarrow (ad)(bc)$
$\mathscr{L}((a(bc))d) =$			$a(bc)d \rightarrow a((bc)d)$
$\mathscr{L}((ab)(cd)) =$			$ab(cd) \rightarrow a(b(cd))$
$\mathscr{L}((a(bd))c) =$			$a(bd)c \rightarrow a((bd)c)$
$\mathscr{L}((ab)cd) =$	$abcd \rightarrow$	$a(bcd)$	

Appendix

Flowcharts

The material of this appendix goes outside the main body of the book for two reasons. First, our object has been to determine programs without regard to whether their instructions would be performed sequentially, or not, but in assigning a flowchart to a program, we are imposing a particular linear ordering on its instructions. The second reason is more fundamental: Our area of investigation is the set of all programs whose instructions belong to a language $\mathscr{C}$ and whose purpose is to list without repetitions the largest possible number of elements of $\mathscr{C}$, on the assumption that the remaining unlistable elements are given. That is, the unlistable elements must be given explicitly or else in the form of a subroutine which lists them by means of instructions which cannot belong to $\mathscr{C}$. The flowcharts, 3 through 12 of this appendix involve such subroutines.

Suppose $\mathscr{C}$ is a programming language, and $\mathscr{P}$ is a program in $\mathscr{C}$ which lists as many elements of $\mathscr{C}$ as possible. $\mathscr{P}$ is defined by giving its input and instructions, all of which belong to $\mathscr{C}$. All other elements of $\mathscr{C}$ are listed by $\mathscr{P}$, and there is no program in $\mathscr{C}$ which can list more elements. The input and instructions, defining $\mathscr{P}$, are given by two kinds of subroutine. One of them is H, which generates the input, and the other is L, which generates for any x in $\mathscr{C}$ the list of all instructions in $\mathscr{C}$ (if any) which depend on x.

Flowcharts 1 and 2 demonstrate how a program $\mathscr{P}$, depends on unspecified subroutines H and L, while flowcharts 3 through 12 describe particular subroutines H and L, which can be substituted into flowcharts 1 and 2.

FLOWCHART 1. Elementary language $\mathscr{C}_1 \cup \mathscr{C}_2$.

Purpose. To list the elements of $\mathscr{C}_1$ without repetitions by means of a program whose instructions are taken from $\mathscr{C}_2$.

Given. The *input* $\mathscr{H}_1$ of the program is given in the form of a partial list of the elements of $\mathscr{C}_1$. The *instructions* of the program are expressed in terms of an operator $\mathscr{L}$, which determines a list $\mathscr{L}(x)$ of those instructions in the program which depend on a known element x of $\mathscr{C}_1$. (When there are instructions which depend on more than one element, $\mathscr{L}(x)$ contains only the instructions such that x in the *last* element of $\mathscr{C}_1$ which the instruction depends on, relative to the linear order in which the program lists the elements of $\mathscr{C}_1$).

Variables. Let $x(1)$, $x(2)$, $x(3)$, ... be variables, to which the program will assign all values in $\mathscr{C}_1$ without repetitions, and let

$$\eta, \lambda, \phi, \psi$$

be variables, to which the program will assign integer values.

START

H_1. This is a subroutine which generates input values and records them thus: $\eta \leftarrow$ number of entries in the list $\mathscr{H}_1$.

$$x(1), x(2), \ldots, x(\eta) \leftarrow \mathscr{H}_1.$$

$\phi, \psi, \leftarrow 1, \eta$

L_1. This is a subroutine which generates output values and records them thus: $\lambda \leftarrow$ number of entries in $\mathscr{L}(x(\phi))$. If $\lambda = 0$, end subroutine; otherwise $x(\psi + 1), x(\psi + 2), \ldots, x(\psi + \lambda) \leftarrow$ output produced by the list $\mathscr{L}(x(\phi))$ of instructions.

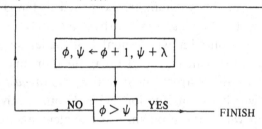

$\phi, \psi \leftarrow \phi + 1, \psi + \lambda$

NO $\quad \phi > \psi \quad$ YES $\quad$ FINISH

Flowchart 1

FLOWCHART 2. Language $\mathscr{C}$.

Purpose. To list the elements of a language,

$$\mathscr{C} = \mathscr{C}_1 \cup \mathscr{C}_2 \cup \ldots \cup \mathscr{C}_\pi \cup \ldots,$$

without repetitions and in non-decreasing order of dimension π by means of a program, whose instructions are elements of the language itself.

Given. The input of the program is given in the form of a sequence

$$\mathscr{H}_1, \mathscr{H}_2, \ldots, \mathscr{H}_\pi, \ldots$$

of lists, where $\mathscr{H}_\pi$ is a partial list of the elements of $\mathscr{C}_\pi$. The *instructions* of the program are expressed in terms of an operator $\mathscr{L}$, which will make a list $\mathscr{L}(x)$ of those instructions in the program which depend on a known element x of the language. (When there are instructions which depend on more than one element, $\mathscr{L}(x)$ contains only the instructions such that x is the *last* element of the language which the instruction depends on, relative to the linear ordering in which the program lists the elements of the language.)

Variables. Let $x(\pi, 1), x(\pi, 2), x(\pi, 3), \ldots$ be variables, to which the program will assign all values from $\mathscr{C}_\pi$ without repetitions, and let

$$\eta, \lambda, \pi, \phi, \psi, \omega$$

be variables, to which the program will assign integer values.

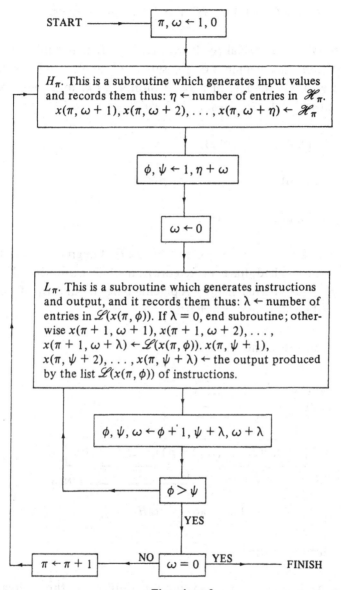

Flowchart 2

FLOWCHART 3. H_1 and L_1 for monomials.

Our objective is to specialize flowchart 1, so that it will list without repetitions the elements of the set $\mathscr{C}_1$ of all monomials which are factors, other than 1, of a given monomial m in μ indeterminates. Let the indeterminates be denoted by

$$a(1), a(2), \ldots, a(\mu),$$

and let their positive integral exponents in m be

$$\rho(1), \rho(2), \ldots, \rho(\mu),$$

respectively. Let the indeterminates or the first degree monomials, as they are also called, have a fixed arbitrary linear ordering. Then, our objective can be reached by replacing parts H_1 and L_1 of flowchart 1 by the following flowcharts:

Part H_1. Input Routine.

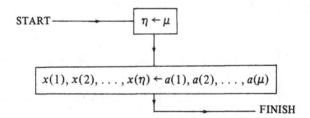

Flowchart 3. Part H_1.

Part L_1. Output routine.

Variables. An integer variable α is used in addition to those already introduced in flowchart 1.

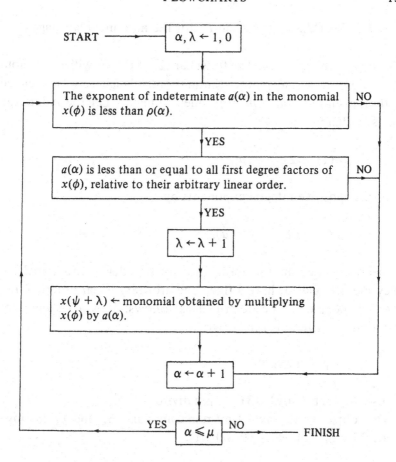

Flowchart 3. Part L_1.

FLOWCHART 4. H_π and L_π for the monomial language.

Our objective is to specialize flowchart 2, so that it will list without repetitions every element of the monomial language $\mathscr{C}$, associated with a given monomial m in μ indeterminates. Let the indeterminates be denoted by

$$a(1), a(2), \ldots, a(\mu)$$

and their positive integral exponents by

$$\rho(1), \rho(2), \ldots, \rho(\mu),$$

respectively. Let the indeterminates, or first degree monomials, as they are also called, have a fixed and arbitrary linear ordering. Let each element of $\mathscr{C}$ be described as follows: It is a sequence of monomials, different from 1, written

$$j(1) \times j(2) \times j(3) \times \ldots,$$

whose product $j(1)j(2)j(3) \ldots$ is a divisor of m.

Our objective is reached by replacing parts H_π and L_π in flowchart 2 by the following flowcharts:

Part H_π. Input Routine.

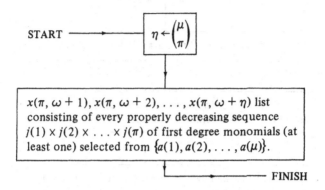

START ──────────────▶ $\eta \leftarrow \binom{\mu}{\pi}$

$x(\pi, \omega + 1), x(\pi, \omega + 2), \ldots, x(\pi, \omega + \eta)$ list consisting of every properly decreasing sequence $j(1) \times j(2) \times \ldots \times j(\pi)$ of first degree monomials (at least one) selected from $\{a(1), a(2), \ldots, a(\mu)\}$.

──────────────── FINISH

Flowchart 4. Part H_π.

N.B. It is not the object of this appendix to give flowcharts which are so explicit that they could be directly translated into computer programs. In the present case a more explicit flowchart would be as follows:

Alternative for Part H_π. An explicit input routine.

Variables. Let $\alpha(1)$, $\alpha(2)$, $\ldots$, $\alpha(\pi)$ be integer variables, used by the program in the expression of any element

$$a(\alpha(1)) \times a(\alpha(2)) \times \ldots \times a(\alpha(\pi))$$

of the input, being generated. Let κ be an integer variable used by the program in addition to those already introduced in flowchart 2.

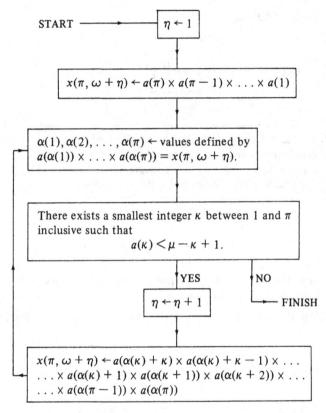

START $\longrightarrow$ $\eta \leftarrow 1$

$x(\pi, \omega + \eta) \leftarrow a(\pi) \times a(\pi - 1) \times \ldots \times a(1)$

$\alpha(1), \alpha(2), \ldots, \alpha(\pi) \leftarrow$ values defined by $a(\alpha(1)) \times \ldots \times a(\alpha(\pi)) = x(\pi, \omega + \eta)$.

There exists a smallest integer κ between 1 and π inclusive such that
$$a(\kappa) < \mu - \kappa + 1.$$

YES   NO

$\eta \leftarrow \eta + 1$ $\longrightarrow$ FINISH

$x(\pi, \omega + \eta) \leftarrow a(\alpha(\kappa) + \kappa) \times a(\alpha(\kappa) + \kappa - 1) \times \ldots$
$\ldots \times a(\alpha(\kappa) + 1) \times a(\alpha(\kappa + 1)) \times a(\alpha(\kappa + 2)) \times \ldots$
$\ldots \times a(\alpha(\pi - 1)) \times a(\alpha(\pi))$

Flowchart 4. Alternative for Part H_π.

Part L_π. Instruction and output routine.

Variables. Integer variables, α and β, are used in addition to the variables already introduced in flowchart 2.

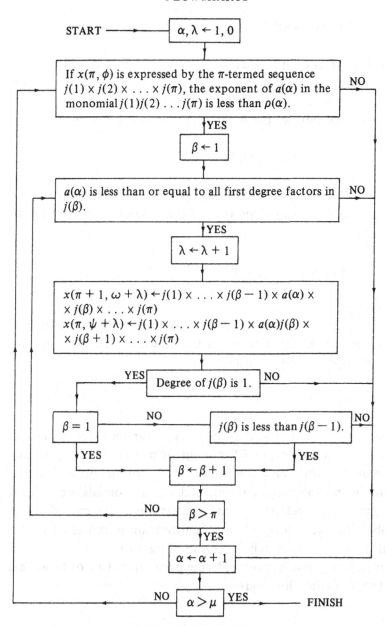

Flowchart 4. Part L_π.

FLOWCHART 5. H_1 and L_1 for factorizations.

Our objective is to specialize flowchart 1, so that it will list without repetitions the elements of the set $\mathscr{C}_1$ of all factorizations of a given monomial m .

In this flowchart let the indeterminates be denoted by

$$a(1), a(2), \ldots, a(\mu),$$

and let this ordering of them be described as their *increasing order*. Let the positive integral powers of these indeterminates in m be denoted by

$$\rho(1), \rho(2), \ldots, \rho(\mu),$$

respectively.

A *factorization* of m is a list of monomials (different from 1) whose product is equal to m . In this flowchart a factorization is uniquely expressed by a sequence

$$(s(1), s(2), \ldots)$$

of monomials, ordered so that their degrees are non-decreasing and so that within an interval of monomials of the same degree they are ordered as follows: Each monomial is written with its indeterminates in non-decreasing order, and the monomials are ordered lexicographically relative to the indeterminates, regarded as an alphabet. In L_1 , below, when a factorization is expressed in this unique manner, is said to be in *conventional form*.

Our object is reached by replacing parts H_1 and L_1 of flowchart 1 by the following flowcharts:

Part H_1. Input routine.

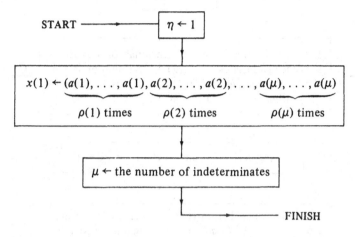

Flowchart 5. Part H_1.

Part L_1. Output routine.

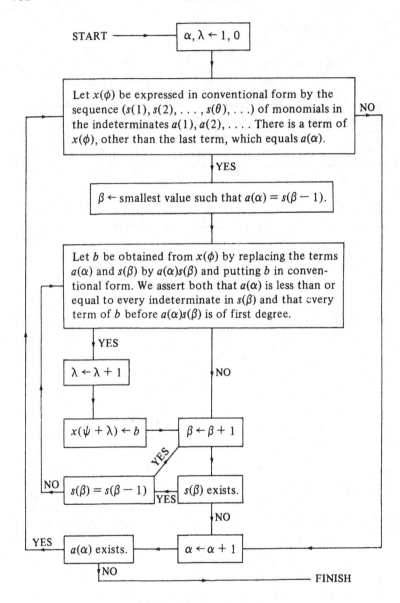

Flowchart 5. Part L_1.

FLOWCHART 6. H_π and L_π for the language of factorizations.

Our objective is to specialize flowchart 2, so that it will list without repetitions every element of the language $\mathscr{C}$ of factorizations of a monomial m, whose indeterminates are

$$a(1), a(2), \ldots , a(\mu),$$

raised to the powers

$$\rho(1), \rho(2), \ldots , \rho(\mu),$$

respectively. Each element of $\mathscr{C}$ is uniquely expressed by a sequence

$$(s(1), s(2), \ldots , s(\theta), \ldots),$$

in which each term $s(\theta)$ is a sequence

$$j(\theta, 1) \times j(\theta, 2) \times \ldots \times j(\theta, \phi) \times \ldots$$

of monomials, such that the product

$$j(\theta, 1)j(\theta, 2) \ldots j(\theta, \phi) \ldots$$

over every θ and ϕ is equal to m. The terms $s(1), s(2), \ldots$ are in non-decreasing order relative to the linear ordering defined in Sections III.3 and III.4. Only odd-dimensional terms can be repeated (i.e. $s(\theta) = s(\theta + 1)$), where the dimension of a term is the number of $j(\theta, \phi)$'s in it. In this flowchart, when the elements of $\mathscr{C}$ are described as being in *conventional form*, it is meant that they are expressed in the manner which was used in the examples at the end of Chapter III.

Our objective is reached by replacing parts H_π and L_π of flowchart 2 by the following flowcharts:

Part H_π. Input routine.

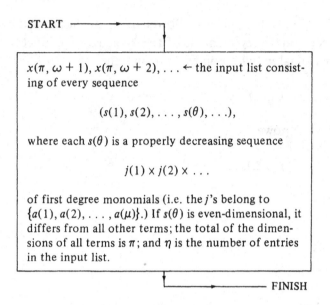

Flowchart 6. Part H_π.

Part L_π. Instruction and output routine.

Variables. Integer variables, α, β, and γ are used in addition to those already introduced in flowchart 2.

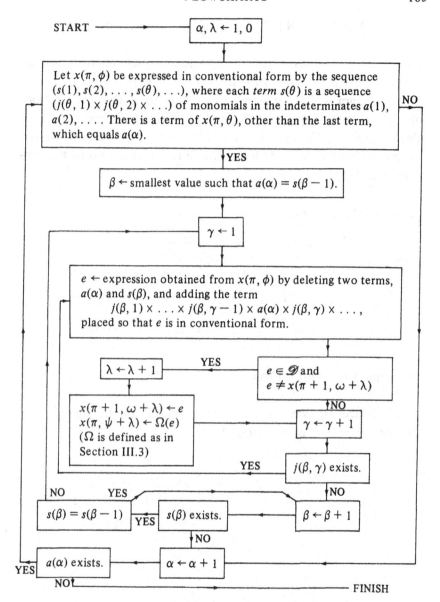

Flowchart 6. Part L_π.

FLOWCHART 7. H_1 and L_1 for transpositions.

Flowchart 7 consists of two components which are to be substituted into parts H_1 and L_1 of flowchart 1. The result is a flowchart for a program which lists without repetitions all the permutations of a given finite sequence. The underlying program was developed algebraically in Chapter IV, and here we introduce a flowchart which specifies a linear order in which the instructions of the program may be performed.

Component H_1. Input routine.

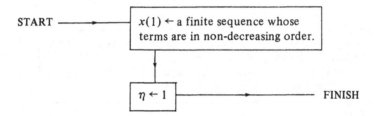

Component L_1. Output routine.

In this flowchart, variables α and y are used in addition to those already introduced in flowchart 1.

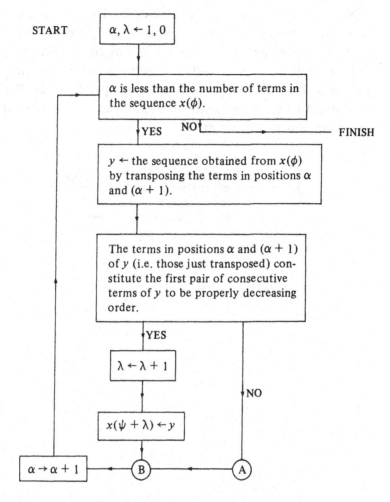

Flowchart 7. Component L_1

Notice that component L_1 satisfies the basic requirement of a flowchart, which is to determine completely the order of performance of instructions, but it is a crude flowchart, because, in general, the path through A will be used more often than necessary. The loop through A does not result in the performance of a transposition. An obvious refinement of L_1 would be to substitute the following step between A and B:

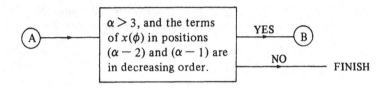

FLOWCHART 8. H_π and L_π for the language of transpositions.

Flowchart 8, consists of two components which are to be substituted for H_π and L_π in flowchart 2. The result is a flowchart for a program which lists without repetitions every element in the language $\mathscr{C}$ of transpositions, where $\mathscr{C}$ is the language in which the program is expressed. The program was developed in Chapter IV, and now it is reduced to a flowchart.

$\mathscr{C}$ can be described as the set of all elements which can be obtained from a given sequence by permuting it and then separating it into intervals called blocks. Each p in $\mathscr{C}$ is a sequence of blocks,

$$p = (l_1, l_2, \ldots, l_\theta, \ldots),$$

in which each block l_θ is itself a sequence, written in multiplicative form

$$l = e_1 \times e_2 \times \ldots,$$

The flowchart makes use of the fact, demonstrated in Section IV.3, that each p in $\mathscr{C}$ has a unique separation into two parts, of which the one on the left is non-void and is called the homological part. It is of a special and simple form.

Part H_π. Input routine.

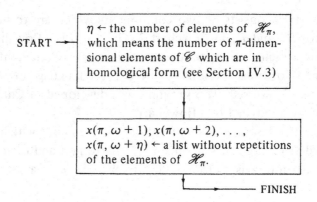

START $\rightarrow$ $\eta \leftarrow$ the number of elements of $\mathscr{H}_\pi$, which means the number of π-dimensional elements of $\mathscr{C}$ which are in homological form (see Section IV.3)

$x(\pi, \omega + 1), x(\pi, \omega + 2), \ldots,$
$x(\pi, \omega + \eta) \leftarrow$ a list without repetitions of the elements of $\mathscr{H}_\pi$.

FINISH

Part L_π. Instruction and output routine.

Variables. The integer variable α is used in addition to those already introduced in flowchart 2.

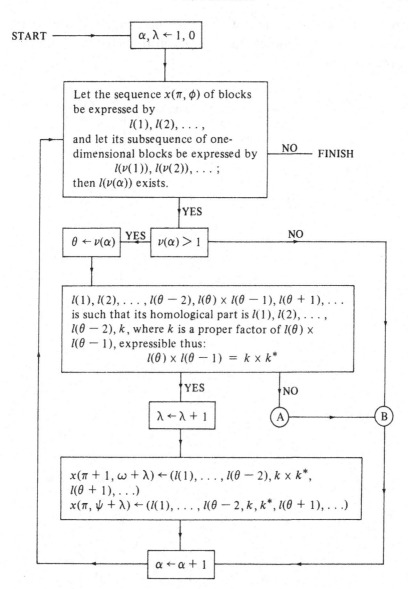

Flowchart 8. Part L_π.

We can take advantage of the fact that the above flowchart examines $x(\pi, \phi)$ from left to right, by introducing the following short-cut between A and B:

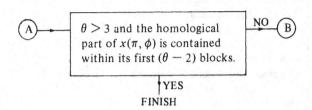

YES

FINISH

FLOWCHART 9. H_1 and L_1 for bracketings

Flowchart 9 consists of two parts, which are to be substituted into parts H_1 and L_1 of flowchart 1. The result specifies a program which lists without repetitions all the complete bracketings of the expression *aaa . . . a* which contains μ appearances of *a*.

Part H_1. Input routine

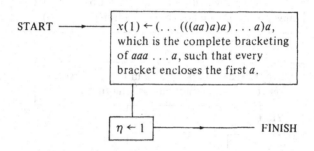

Flowchart 9. Part H_1.

Part L_1. Output routine

This flowchart uses variables α, β, and y, as well as those already introduced in flowchart 1.

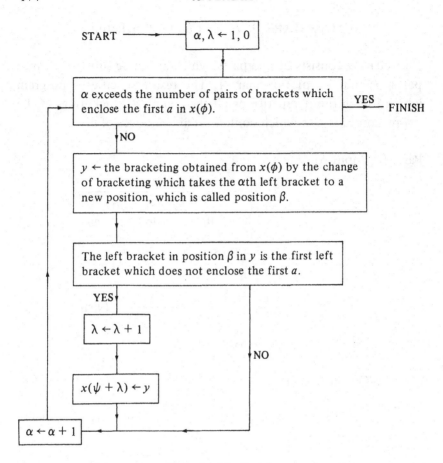

Flowchart 9. Part L_1.

FLOWCHART 10. H_π and L_π for the language of bracketings.

Flowchart 10 consists of two parts, which are to be substituted for H_π and L_π in flowchart 2. The result is a flowchart for a program which lists without repetitions every element of the language $\mathscr{C}$ of bracketings, where $\mathscr{C}$ is the language in which the program is expressed. The program was developed in Chapter V, and now it is to be reduced to a flowchart.

As an abstract set $\mathscr{C}$ is simply all ways of putting pairs of brackets into the sequence $aaa \ldots a$, in which a appears μ times.

Part H_π. Input routine.

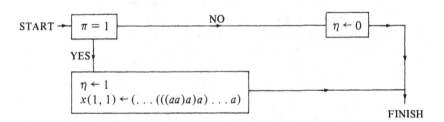

Flowchart 10. Part H_π.

Part L_π. Instruction and output routine.

Variables. The variables α and y are used in addition to those already introduced in flowchart 2.

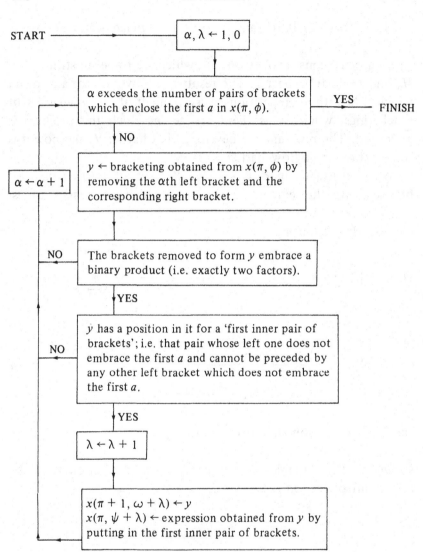

Flowchart 10. Part L_π.

FLOWCHART 11. H_1 and L_1 for Clusterings.

Flowchart 11 consists of two parts, which are to be substituted into Parts H_1 and L_1 of flowchart 1. The result specifies a program which lists without repetitions all the complete clusterings of the set

$$\{a_1 a_2, \ldots, a_\mu\},$$

expressed as bracketed sequences. The instructions, used in this program, shift brackets around, but do not permute letters. Consequently, the input must include a whole set of permutations.

Part H_1. Input routine.

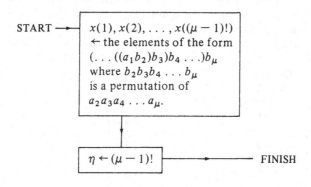

Flowchart 11. Part H_1.

Part L_1. Output routine.

This flowchart uses variables α, β, y, as well as those already introduced in flowchart 1.

START

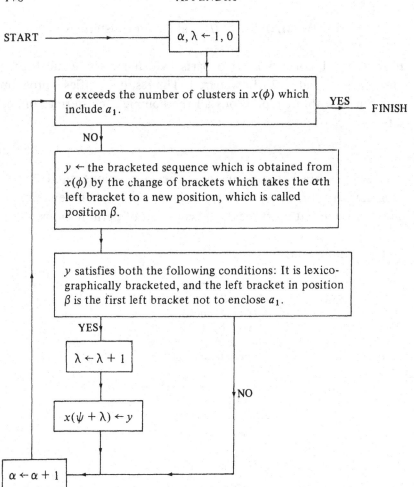

Flowchart 11. Part L_1.

FLOWCHART 12.

H_π and L_π for the language of clusterings of a set.

Flowchart 12 consists of two parts, which are to be substituted for H_π and L_π in flowchart 2. The result is a flowchart for a program which lists without repetitions every element of the language $\mathscr{C}$ of clusterings, where $\mathscr{C}$ is the language in which the program is expressed. The program was developed in Chapter VI, and now it is to be reduced to a flowchart.

As an abstract set $\mathscr{C}$ is simply all the ways of reducing the set $\{a_1, a_2, \ldots, a_\mu\}$ to a family of clusters, a cluster being a proper subset with more than one element. Each clustering is expressed by a lexicographically bracketed sequence, as defined at the end of Section VI.3.

Part H_π. Input routine.

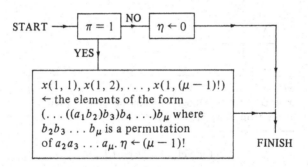

Flowchart 12. Part H_π.

Part L_π. Instruction and output routine.

Variables. The variables α, y, and z are used in addition to those already introduced in flowchart 2.

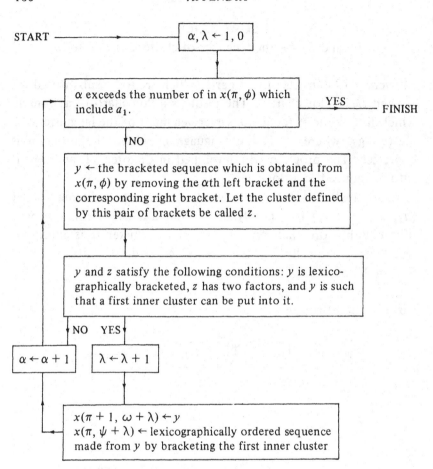

START ⟶ $\alpha, \lambda \leftarrow 1, 0$

α exceeds the number of in $x(\pi, \phi)$ which include a_1. —— YES —— FINISH

NO

$y \leftarrow$ the bracketed sequence which is obtained from $x(\pi, \phi)$ by removing the αth left bracket and the corresponding right bracket. Let the cluster defined by this pair of brackets be called z.

y and z satisfy the following conditions: y is lexicographically bracketed, z has two factors, and y is such that a first inner cluster can be put into it.

NO   YES

$\alpha \leftarrow \alpha + 1$ $\lambda \leftarrow \lambda + 1$

$x(\pi + 1, \omega + \lambda) \leftarrow y$
$x(\pi, \psi + \lambda) \leftarrow$ lexicographically ordered sequence made from y by bracketing the first inner cluster

Flowchart 12. Part L_π.

References

1. Even, Shimon, *Algorithmic Combinatorics*, The Macmillan Company, New York, 1973.
2. Foata, Dominique, 'Etude algébrique de certains problèmes d'analyse combinatoire et du calcule des probabilités', *Publ. Inst. Statist. Univ.*, Paris 14 (1965), p. 176.
3. Hilton, P. J., and Wylie, S., *Homology Theory*, Cambridge University Press, 1962.
4. Knuth, Donald E., *The Art of Computer Programming*, Addison-Wesley Publishing Company, Reading, Mass., 1968.
5. Nijenhuis, A., and Wilf, H. S., *Combinatorial Algorithms*, Academic Press, New York, 1975.

Index